Microbiology

PEARLS OF WISDOM

Related Titles in Microbiology

Stay in the loop! Sign up for Jones and Bartlett's eUpdates today and receive periodic emails alerting you to the latest texts and resources in the biological sciences.
Sign up today at www.jbpub.com/email.

Alcamo's Fundamentals of Microbiology, Eighth Edition, Pommerville

Alcamo's Laboratory Fundamentals of Microbiology, Eighth Edition, Pommerville

Alcamo's Fundamentals of Microbiology: Body Systems Edition, Pommerville

Guide to Infectious Diseases by Body System, Pommerville

Encounters in Microbiology, Volume One, Second Edition, Pommerville

Encounters in Microbiology, Volume Two, Pommerville

Principles of Modern Microbiology, Wheelis

Microbes and Society, Second Edition, Weeks and Alcamo

The Microbial Challenge: Human-Microbe Interactions, Krasner

Understanding Viruses, Shors

AIDS: The Biological Basis, Fifth Edition, Weeks and Alcamo

AIDS: Science and Society, Fifth Edition, Fan, Conner and Villarreal

20th Century Microbe Hunters, Krasner

How Pathogenic Viruses Work, Sompayrac

Human Parasitic Diseases Sourcebook, Berger and Marr

Protoctista Glossary, Margulis

Electron Microscopy, Second Edition, Bozzola and Russell

Other Titles in Jones and Bartlett's Pearls of Wisdom Series

EMT-Basic: Pearls of Wisdom, Second Edition, Haskell

EMT-Intermediate: Pearls of Wisdom, Second Edition, Haskell

General Chemistry: Pearls of Wisdom, Clark

General Physics: Pearls of Wisdom, Amstutz

General Surgery Medical Student USMLE Parts II and III: Pearls of Wisdom, Breckler

Internal Medicine Medical Student USMLE Parts II and III: Pearls of Wisdom, Zevitz

MCAT Review: Pearls of Wisdom, Second Edition, Donnino

Medical Biochemistry: Pearls of Wisdom, Eichler

Organic Chemistry: Pearls of Wisdom, Vallombroso

Paramedic: Pearls of Wisdom, Second Edition, Haskell

Pathology Medical Student USMLE Part I: Pearls of Wisdom, Roberts

Pediatric Medical Student USMLE Parts II and III: Pearls of Wisdom, Levin

Pediatric Nurse Practitioner: Pearls of Wisdom, Kriegler

Pharmacology: Pearls of Wisdom, Scholar

The Law of Contracts: Pearls of Wisdom, Denicola

JONES AND BARTLETT'S
PEARLS OF WISDOM SERIES

Microbiology

PEARLS OF WISDOM

Second Edition

S. James Booth

Department of Pathology and Microbiology
University of Nebraska Medical Center

JONES AND BARTLETT PUBLISHERS

Sudbury, Massachusetts

BOSTON   TORONTO   LONDON   SINGAPORE

Jones and Bartlett Publishers
World Headquarters
Jones and Bartlett Publishers
40 Tall Pine Drive
Sudbury, MA 01776
978-443-5000
info@jbpub.com
www.jbpub.com

Jones and Bartlett Publishers Canada
6339 Ormindale Way
Mississauga, Ontario L5V 1J2
Canada

Jones and Bartlett Publishers International
Barb House, Barb Mews
London W6 7PA
United Kingdom

Jones and Bartlett's books and products are available through most bookstores and online booksellers. To contact Jones and Bartlett Publishers directly, call 800-832-0034, fax 978-443-8000, or visit our website, www.jbpub.com.

Production Credits
Chief Executive Officer: Clayton Jones
Chief Operating Officer: Don W. Jones, Jr.
President, Higher Education and Professional Publishing: Robert W. Holland, Jr.
V.P., Sales and Marketing: William J. Kane
V.P., Design and Production: Anne Spencer
V.P., Manufacturing and Inventory Control: Therese Connell
Publisher, Higher Education: Cathleen Sether
Acquisitions Editor, Science: Molly Steinbach
Managing Editor, Science: Dean W. DeChambeau
Associate Editor, Science: Megan R. Turner
Editorial Assistant, Science: Caroline Perry
Production Manager: Louis C. Bruno, Jr.
Associate Production Editor: Leah Corrigan
Senior Marketing Manager: Andrea DeFronzo
Composition: SNP Best-set Typesetter Ltd.
Cover Design: Kate Ternullo
Printing and Binding: Courier Stoughton
Cover Printing: Courier Stoughton

Library of Congress Cataloging-in-Publication Data
Booth, S. James.
    Microbiology : pearls of wisdom / S. James Booth. – 2nd ed.
        p. ; cm. – (Pearls of wisdom)
    ISBN 978-0-7637-6860-7
    1. Microbiology–Miscellanea.   I. Title.   II. Series: Pearls of wisdom series (Sudbury, Mass.)
    [DNLM: 1. Microbiology–Examination Questions. QW 18.2 B725m 2010]
    QR62.B66 2010
    579–dc22
                                                            2008041379

6048

Printed in the United States of America
13 12 11 10 09   10 9 8 7 6 5 4 3 2 1

Table of Contents

Preface

The approximately two thousand revised and updated questions in *Microbiology: Pearls of Wisdom, Second Edition* will help you pass medical microbiology and improve your board scores. The primary intent of *Pearls* is to serve as a rapid review of microbiology principles, particularly as they apply to medical microbiology, and to serve as a study aid to improve your performance on medical microbiology examinations. With this goal in mind, the text is written in a unique rapid-fire, question/answer format that is more efficient than the multiple choice format common in most other review books. You will receive immediate gratification with a correct answer. Questions themselves often contain a "pearl" reinforced in association with the question/answer.

Additional hooks are often attached to the answer in various forms, including mnemonics, visual imagery, repetition, and humor. Information not requested in the question is often included in the answer and serves as a nexus to help in retrieving the information during an exam. Information is often included in more than one question and answer, worded differently, but accurately, to help those with alternative learning styles. Emphasis has been placed on evoking both key and "trivial" facts that are easily overlooked or quickly forgotten and yet somehow always seem to appear on microbiology exams. Remember, what you might consider trivia, a microbiologist might consider essential knowledge. Some questions have answers without explanations. This is done to enhance the ease of reading and rate of learning. Explanations often occur in a later question or answer.

It may happen that upon reading an answer you may think, "Why is that?" or, "Is that right?" If this happens to you, GO CHECK! While researching answers to questions, I usually checked several current sources, just to make sure. The answers given to questions might differ from what you think is correct for two main reasons: (1) the answers may change as new research becomes available, or (2) your professors may not be keeping up with the current knowledge (no fault of theirs; it's a broad field). To truly assimilate these disparate facts into a framework of knowledge absolutely requires further reading in the surrounding concepts. Information learned by actively seeking an answer to a particular question is much better retained than information passively read. And be sure to take notes as you progress through this book! If you have a book from which you are trying to learn and you do not write notes in the book, on a sheet of paper, in a word processor, etc., you are using the book incorrectly and not learning as much as you could.

To best use this book, use a 3 × 5 card to cover the answers. If you miss the answer, put a check beside it using blue or black ink. After you complete a section, start over. Skip the questions that you got correct the first time and only attempt the questions with a check. If you miss a checked question a second time, place another check using red ink. After you complete the section again, start over. Only attempt the questions with the red checks (red makes them easier to find quickly, and time is at a premium). If you miss the question again (i.e., this is your third attempt), forget the question. You will probably miss the question on the exam. However, if you are stubborn you can try to learn the material, get the answer correct on the exam, and prove me wrong!

Pearls does have limitations. By its very nature, soon after publication, many of the concepts will not represent the cutting edge of microbiology. There may be conflicts among texts and even among courses of medical microbiology. New discoveries may result in knowledge that deviates from that which represents the correct answer on an examination. To minimize this, I have selected the information I believe is most likely correct for test purposes. As I tell my own students, there are exceptions to almost everything that is taught to you. Fortunately, most exams are written to see if you know and understand the general principles, not these minor exceptions.

Great effort has been made to verify the answers to the questions. I have updated the numerous changes in the names of organisms since the first edition. Drugs of choice were current at publication, but will likely change as drug resistance emerges and new therapies evolve. Some questions retained from the first edition now have different answers reflecting new discoveries. Some of these may change again. As stated earlier, if you doubt an answer, GO CHECK! Please make me aware of any errors or oversights you find; I welcome your comments, suggestions, and criticisms. I hope to make improvements in a third edition and would greatly appreciate your input. Please write to S. James Booth, PhD at jbooth@unmc.edu. I look forward to hearing from you.

Study hard and good luck!

S.J.B.

To Jodi, Mike, Jason, Nathan, and Stephanie, who have always brought joy to my life;
and to the students at the University of Nebraska Medical Center
whose positive comments have encouraged me to write this second edition.

Basic Bacteriology

Q: A 7-year-old child with impetigo visited her pediatrician with her father. Pus from one of the blister-like lesions was collected and Gram stained. Many gram-positive cocci in irregular clusters were observed. What is the most likely pathogen?

A: *Staphylococcus aureus*

Q: Blood from a patient with a postsurgical abdominal infection was submitted to the laboratory for culture. The isolated organism would not grow in the presence of oxygen. Subsequently, it was shown to produce endospores. What is the genus of this organism?

A: *Clostridium*, an anaerobe

Q: With respect to the previous question, what is the genus if the organism could grow in the presence of oxygen?

A: *Bacillus*, but it is unlikely in the above scenario. Some species of *Bacillus* are obligate aerobes, whereas others are facultative. For convenience, we usually say they are aerobes. Note: In medical microbiology, we typically consider only the genera of *Bacillus* and *Clostridium*. As it turns out there are other genera of spore-formers such as *Sporolactobacillus*, *Paenibacillus*, *Brevibacillus*, *Aneurinibacillus*, *Geobacillus*, and *Virgibacillus*. Fortunately, you do not need to know any of these because they rarely (if ever) cause human infections.

Q: What are the major differences between the cell walls of gram-positive and gram-negative bacteria?

A: Gram-positive bacterial cell walls are composed in large part of a thick peptidoglycan outside of the cytoplasmic membrane. Gram-negative bacterial cell walls consist of a cytoplasmic, or inner, membrane and an outer membrane separated by a periplasmic space where a thin layer of peptidoglycan is found.

Q: List at least three *facultative* intracellular bacteria.

A: Seven of the most important follow: *Listeria monocytogenes*, *Mycobacterium tuberculosis*, *Salmonella*, *Shigella*, *Legionella pneumophila*, *Brucella*, and *Yersinia pestis*. The first five are the most important to remember as facultative intracellular parasites. There are other reasons to know about *Brucella* and *Y. pestis*. This is not a complete list; there are others. Facultative intracellular bacteria are those that grow both on laboratory media and within certain cell types, often within macrophages. Obligate intracellular parasites, such as *Chlamydia*, *Chlamydophila*, *Rickettsia*, *Coxiella*, *Ehrlichia*, and *Anaplasma* (formerly *Ehrlichia*), cannot grow on laboratory media. Note that just because an organism cannot grow on laboratory media (e.g., *Treponema pallidum*, the cause of syphilis) does not necessarily mean it is an obligate intracellular parasite.

Q: What is the single most important activity that decreases the nosocomial infection rate?

A: Hand washing; however, many nosocomial infections are due to one's own normal flora. You might argue that other interventions, such as decontamination of skin before surgery or before insertion of a catheter, are more important, but medical personnel already do these, so they would have little effect on reducing the nosocomial infection rate any further.

Q: What name is given to the phenomenon by which *Neisseria gonorrhoeae* can evade the immune system?

A: Antigenic variation or phase variation

Q: Antigenic or phase variation is related to what structures on *Neisseria gonorrhoeae*?

A: Pili, primarily, but also various cell wall proteins. With respect to pili, antigenic variation relates to modifications in the amino acid sequences of the hypervariable region of the pili protein. In some patients, antigenic variation results in reinfection by the same organism that has "new" pili for which the patient has no protective antibodies against. For your purposes, antigenic and phase variation should be thought of as being synonymous. Microbiologists, however, distinguish between the two. Antigenic variation refers to a modification in the amino acid sequence of a (usually) structural protein. The structure is still there, but the protein that makes it has been slightly modified. Phase variation relates to the expression or nonexpression (like an on–off switch) of bacterial pili (or other proteins). Nonpiliated strains are avirulent. Antigenic and phase variation are not unique to *Neisseria*. They are found in a large number of pathogenic bacteria (e.g., *Borrelia recurrentis*, *Bordetella pertussis*, *Escherichia coli*, *Haemophilus influenzae*, *Helicobacter pylori*, *Streptococcus pneumoniae*, etc.) and with some parasitic diseases (e.g., malaria and African trypanosomiasis).

Q: What is the name of the outer leaflet of the outer membrane of gram-negative bacteria?

A: Lipopolysaccharide (LPS) or endotoxin

Q: Describe the structure of the LPS of gram-negative bacteria.

A: LPS is composed of lipid A and a covalently attached core polysaccharide. The lipid A component is the biologically active toxic portion. The terminal repeat units attached to the core polysaccharide comprise the O antigen, which is partly responsible for the antigenic specificity of some gram-negative bacteria. Note: Some gram-negative bacteria, such as *Neisseria*, *Haemophilus*, and *Campylobacter*, contain LOS, or lipo-oligosaccharide, rather than LPS. LOS is analogous to LPS. LOS, however, does not have the O antigens found in LPS. For standardized exams, you do not need to know the difference. Just recognize that LOS is the endotoxin of *Neisseria* and several other gram-negative bacteria.

Q: List some biological activities of endotoxin.

A: Two of the most common effects of endotoxin are fever production and shock. Other effects include activation of complement, B-cell stimulation, leukopenia, and disseminated intravascular coagulation. The effects of endotoxin are due to the release of cytokines such as interleukin-1 (IL-1), IL-6, IL-8, tumor necrosis factor alpha, and interferon gamma. The list of mediators is large, thus leading to a diverse array of endotoxin-mediated biological activities.

Q: What is the role of Toll-like receptors in innate immunity?

A: Toll-like receptors, or TLRs, are a type of cell-associated proteins called pattern recognition receptors, or PRRs. PRRs are germline-encoded proteins in mammals that function to recognize pathogen-associated molecular patterns, or PAMPs. The PAMPs are molecular structures specific to microbes. An example is the LPS of gram-negative bacteria. PRRs such as TLRs (and others, such as mannose-binding protein) activate immune cells, causing an induction of a variety of genes involved in innate and adaptive immunity. For example, PRRs activate the complement pathway (innate immunity) and can induce the production of a variety of cytokines (e.g., IL-1, IL-6, and tumor necrosis factor) and other immune mediators (e.g., chemokines) to induce the inflammatory response, recruit neutrophils to the infection site, and activate macrophages to phagocytize and kill the pathogens.

Q: How do bacteria release exotoxins and endotoxins (what is the mechanism)?

A: Exotoxins are typically secreted. Gram-negative bacterial cell walls are more complex than those of gram-positive bacteria and require sophisticated bacterial secretion systems. One classification scheme includes six of these systems, designating each as types I through VI. Types III and IV are particularly important in bacterial virulence. Endotoxins are released after cell lysis of the gram-negative bacteria. They are not present in gram-positive cell walls.

Q: An environmental microbiologist collected a water sample from an air conditioning cooling tower. One of the organisms found living in the water was viewed under a dark-field microscope and was observed to have a distinct nucleus, including a nuclear membrane. Would this organism be expected to be susceptible or resistant to penicillin?

A: It is resistant to penicillin. The description given is for a eukaryotic organism (e.g., amoeba, etc.), not prokaryotic (bacteria). Penicillin (a beta-lactam antibiotic) inhibits peptidoglycan synthesis in the cell walls of susceptible bacteria.

Q: What name is given to proteins secreted by some bacteria that chelate iron?

A: Siderophores

Q: What term is used to refer to microbes that normally cause disease only in immunocompromised people?

A: Opportunistic pathogens

Q: In 1884 a German physician developed a set of "rules" that could be used to help determine if a specific microorganism was the cause of a specific disease. What is the name of this set of rules?

A: Koch's postulates. These rules, developed by the Nobel Prize-winning German microbiologist Robert Koch are as follows: (1) The microorganism should be found in all cases, (2) the microorganism should be isolated and grown in pure culture, (3) the isolated microorganism should cause the same disease when inoculated into a susceptible animal, and (4) the microorganism should be reisolated from the experimental infection. Some diseases may not satisfy all of Koch's postulates. For example, *Treponema pallidum*, the causative agent of syphilis, cannot be grown in the lab on artificial media.

Q: Our normal microbial flora can function to prevent infection by other bacteria. How?
  A: It helps to prevent colonization with potential pathogens that cannot compete well with our normal flora. In addition, many species of our normal flora produce substances such as fatty acids and other antibacterial compounds, including bacteriocins, that are inhibitory or kill other bacteria.

Q: *Escherichia coli* is an important cause of urinary tract infections, gastroenteritis, and sepsis. One of its virulence factors has been referred to as a K antigen. What is the K antigen?
  A: A capsule

Q: What bacteria are usually being referred to with the following descriptions?

a. Kidney bean shaped
b. Lancet shaped
c. Seagull appearance
d. Swarming
e. Blue-green pigment
f. Chinese letters
g. Box car shaped
h. Safety pin shaped

  A: These handy descriptive terms can help narrow down the possibilities, but make sure they fit the clinical picture before you make some costly assumptions.

  a. *Neisseria* is a best guess, but it could be *Moraxella*.
  b. *Streptococcus pneumoniae*
  c. *Campylobacter*
  d. *Proteus*
  e. *Pseudomonas aeruginosa*
  f. *Corynebacterium*; it can also look like Y's and V's. *Listeria* resembles *Corynebacterium*.
  g. *Clostridium perfringens* or *Bacillus anthracis*
  h. *Yersinia pestis*

The next five questions are based on Figure 1-1.

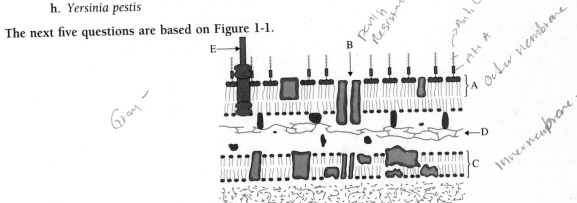

Figure 1-1

Q: Is this a cross-section of a gram-positive or a gram-negative cell wall?
  A: Gram-negative. Gram-positive bacteria do not have an outer membrane.

Q: Removal of which structure (A, B, C, D, or E) would result in a spheroplast?
  A: D, the peptidoglycan layer. Spheroplasts are gram-negative bacteria that have no peptidoglycan. Protoplasts are gram-positive bacteria that have no peptidoglycan.

Q: Which structure (A, B, C, D, or E) is responsible for natural resistance to vancomycin?
  A: B, the porin. The aqueous-filled pore of the porin is too small to allow the large molecular weight (about 1400 daltons) vancomycin to pass through.

Q: Which structure (A, B, C, D, or E) is toxic to humans if the bacteria lyse?

A: A, the endotoxin. Gram-positive bacteria have no endotoxin.

Q: Which structure (A, B, C, D, or E) is responsible for the ability of *Salmonella* to deliver a toxin across the eukaryotic cellular membrane and into the cytoplasm of an intestinal mucosal cell?

A: E, the type III secretion system. Many gram-negative bacteria have specialized secretion systems (types I–VI, so far), allowing them to secrete proteins, including toxins, from their cytoplasm across their cytoplasmic membranes, periplasmic space, and outer membranes to the outside of the cell. These are important virulence factors for these bacteria. The type III secretion system is one that allows bacteria to deliver these protein effectors across eukaryotic cellular membranes directly into their cytoplasm.

Q: A third-year medical student was asked to make a Gram stain of a needle aspirate from the distal femur of a 9-year-old girl with confirmed osteomyelitis. When the medical student viewed the organism under oil, she observed gram-positive cocci in irregular clusters. Assuming that the organism in the infection was the most common one generally found in cases of osteomyelitis, what mistake, if any, did she most likely make in the Gram stain procedure?

A: She made no mistake. The most common organism is *Staphylococcus aureus* (gram-positive cocci in irregular [grape-like] clusters).

Q: Both lysozyme and penicillin have an effect on the cell walls of bacteria. What part of the cell wall is affected?

A: They both have an effect on the peptidoglycan.

Q: What specifically is the effect of lysozyme on the peptidoglycan (i.e., what is the specific target and what happens to the cell)?

A: Lysozyme breaks the beta 1–4 bond between the N-acetyl-glucosamine and the N-acetyl-muramic acid of peptidoglycan, resulting in cell death by lysis (in a hypertonic environment).

Q: What specifically is the effect of penicillin on the peptidoglycan (i.e., what is the specific target and what happens to the cell)?

A: Penicillin inhibits the bacterial enzyme transpeptidase. Transpeptidase is responsible for cross-linking the amino acid side chains of peptidoglycan to each other. Susceptible cells die by lysis (in a hypertonic environment). Notice that lysozyme breaks down existing peptidoglycan and penicillin stops its synthesis. Another name for transpeptidase is "penicillin binding protein." Do not confuse this bacterial enzyme with the liver enzyme gamma glutamyl transpeptidase.

Q: During a senior elective in a research laboratory, a fourth-year medical student was growing a thermophilic bacterium in a culture medium. The generation time was determined to be 15 minutes. If the initial inoculum (i.e., starting number of cells) was one million (1×10^6), how many cells would there be after 1.5 hours? Assume no lag phase (i.e., the cells are in log phase the entire 90 minutes).

A: 64×10^6. 0 time = 1×10^6, 15 minutes = 2×10^6, etc.

Q: Are mesosomes found in gram-positive or gram-negative bacteria?

A: Gram positive. They are invaginations of the cytoplasmic membrane and can only be seen by electron microscopy. Mesosomes are not important for you to know. This question is here because my first seminar as a graduate student was on mesosomes and therefore they are somewhat nostalgic for me.

Q: What is another name for the repeat unit (polysaccharide) of gram-negative bacteria?

A: O antigen

Q: The Vi (virulence) antigen of *Salmonella* refers to what structure?

A: Capsule. Other bacteria (e.g., *E. coli, Citrobacter*) have Vi antigens, but it is most commonly associated with *Salmonella typhi*. It is thought to protect bacteria from phagocytosis, thus making them more virulent. The Vi antigen physically covers the O antigen, protecting it from anti-O antibodies.

Q: In an experiment measuring bacterial growth, *Clostridium perfringens* cells were grown on glucose as the sole carbon and energy source. Subsequently, the bacteria were transferred to a different growth medium containing only chondroitin sulfate as the sole carbon and energy source. On initial transfer to this new medium, the *C. perfringens* would most likely enter which growth phase?

A: Lag phase. *C. perfringens* needs time to adjust to the new substrate by synthesizing inducible enzymes that will degrade the chondroitin sulfate.

Q: What are bacteria called that can grow at high salt concentrations?

    A: Halophiles

Q: What name is given to the region of a bacterial cell where the DNA is located?

    A: Nucleoid. There is no nuclear membrane in bacteria.

Q: What process is used by aerobic bacteria for energy production?

    A: Respiration

Q: What are hemolysins?

    A: Proteins that cause the lysis of red blood cells. They are detected by observing their effect on the erythrocytes, commonly sheep blood, in blood agar plates; hence, we call them hemolysins. However, they are commonly cytotoxic to other types of cells. Their mechanisms for damaging membranes include direct enzymatic damage, forming pores (various cytolysins), or surfactant activity. Many of the various toxins produced by bacteria have hemolytic activity.

Q: List and describe the three hemolytic patterns commonly observed on blood agar plates.

    A: Nonhemolysis (unfortunately, commonly referred to as gamma hemolysis; blood remains red, no clearing or color change), alpha hemolysis (partial or incomplete hemolysis that turns blood in plate a greenish color), and beta hemolysis (complete hemolysis that causes all the red blood cells to lyse, revealing the normal clear yellow coloration of the agar medium).

       Trivia: There are no enzymes or toxins that are responsible for the alpha hemolysis of some streptococci or other bacteria. The greening or partial destruction of red blood cells is due to the production of hydrogen peroxide by the streptococci that lyse some of the red blood cells. This releases hemoglobin into the medium and appears green. If oxygen is removed from the growth atmosphere, peroxide is not formed, and cultures that appeared "alpha hemolytic" now appear nonhemolytic.

Q: What process is used by anaerobic bacteria for energy production?

    A: Fermentation

Q: What is the terminal electron acceptor in aerobic respiration?

    A: Oxygen

Q: What is the terminal electron acceptor in *anaerobic* respiration?

    A: Typically, an inorganic ion other than oxygen, such as nitrate or sulfate.

Q: What group of bacteria naturally lacks cell walls?

    A: Mycoplasmas

Q: What genera of pathogenic bacteria produce endospores?

    A: *Bacillus* and *Clostridium*

Q: What are the oxygen requirements for the two bacteria in the previous question?

    A: *Bacillus* species are aerobes; *Clostridium* species are anaerobes.

Q: What color are gram-positive and gram-negative bacteria after Gram staining?

    A: Gram-positive bacteria are blue; gram-negative bacteria are red or pink.

Q: What is the name of organisms that require organic compounds for growth?

    A: Heterotrophs. It is unlikely this question will occur on a standardized exam.

Q: There are two basic kinds of electron microscopes. Which kind results in a three-dimensional image?

    A: Scanning electron microscope, or SEM. The other basic kind is the transmission electron microscope, or TEM.

Q: What two types of microscopy can be used to view living cells?

    A: Dark-field microscopy and phase contrast microscopy

Q: List some structural differences between eukaryotic and prokaryotic cells.

    A: **Eukaryotic:** Nuclear membrane, multiple chromosomes of linear DNA, 80S ribosomes, endoplasmic reticulum, Golgi apparatus, mitochondria (and chloroplasts if photosynthetic), microtubules, simple cell wall (if cell wall is present), complex flagella or cilia (if present)

    Prokaryotic: Single chromosome composed of circular DNA (usually, however, *Borrelia burgdorferi* and some *Streptomyces* have linear DNA) in nucleoid without nuclear membrane, 70S ribosomes, most have a complex cell wall containing peptidoglycan (except the mycoplasmas). You should be familiar with the major differences between the cell walls of gram-positive and gram-negative bacteria.

Q: Metachromatic granules (also called volutin) are present in what important pathogen?

A: *Corynebacterium diphtheriae* is the usual answer, but they can be found in other corynebacteria and other genera, including *Mycobacterium*, and in some yeasts and parasites.

Q: What are metachromatic granules composed of and how are they demonstrated?

A: They are storage granules of inorganic phosphate and can be demonstrated by staining with methylene blue and viewing the resultant red granules by bright-field microscopy.

Q: Which would be most susceptible to lysozyme, gram-positive or gram-negative bacteria?

A: Gram-positive, because their outer cell wall is composed in large part of peptidoglycan (the target for lysozyme). Lysozyme is too large to get through outer membrane porins of gram-negative bacteria. If the outer membrane is damaged, such as within a white blood cell phagosome, the lysozyme can degrade the peptidoglycan.

Q: What is the major function of pili?

A: Adherence, although sex pili function in conjugation. The pili of gram-negative bacteria are commonly major virulence factors for pathogens, functioning in adherence to tissues.

Q: What substances are responsible for the acid-fast staining reaction of *Mycobacterium*?

A: Waxes and lipids in the cell walls

Q: Where are teichoic acids found?

A: Cell walls of gram-positive bacteria. Although they are structural components of cell walls, they may also function in virulence. For example, they appear to contribute to pharyngeal epithelial cell attachment by *Streptococcus pyogenes*.

Q: What are probiotics?

A: Live microbes that when administered in adequate amounts to a host, confer a health benefit on the host. The most common examples are the bacteria, such as *Lactobacillus acidophilus*, in yogurt and acidophilous milk. Some yeast species have also been used as probiotics. They have been suggested for use (frequently successfully) in diseases such as diarrhea, *Clostridium difficile* colitis, irritable bowel syndrome, urogenital infections, and atopic eczema.

Q: What is the major composition of most bacterial capsules?

A: Polysaccharide, but the capsule of *Bacillus anthracis* is polypeptide (polyglutamic acid)

Q: What name is given to the cells after the removal of the cell wall (peptidoglycan) from gram-positive bacteria?

A: Protoplasts

Q: Removal of the cell wall, specifically the peptidoglycan, of gram-negative bacteria results in what?

A: Spheroplasts, which have sections of outer membrane remaining attached to the cytoplasmic membrane

Q: What is the area called between the outer membrane and the cytoplasmic membrane of gram-negative bacteria?

A: Periplasmic space

Q: What is the name of the cell wall extension that grows inward to eventually produce two daughter cells?

A: Septum

Q: List some factors that lead to sporulation by those bacteria capable of sporulation.

A: Nutrient depletion or other harsh environmental conditions such as accumulation of metabolic byproducts. Sporulation, at least in some species, appears to be controlled by quorum sensing mechanisms. (There is a question later concerning quorum sensing.)

Q: List some structural features of bacterial endospores.

A: Core consisting of single chromosome and enzymes required for germination; high calcium content (bound to dipicolinic acid); little water content; several layers of spore wall, including the cortex (highly cross-linked peptidoglycan) and the spore coat (keratin-like protein)

Q: What are heterotrophic bacteria?

A: Bacteria that require organic compounds as a source for both their carbon and energy. Virtually all pathogenic bacteria are heterotrophs.

Q: Differentiate between anabolism and catabolism.

A: Anabolism is the process of cell synthesis (e.g., amino acid or pyrimidine synthesis). Catabolism is the process of breaking down substrates into small molecules that can be used in cell synthesis (anabolism). Catabolism also yields energy that can be used in anabolic processes.

Q: Glucose is converted into what final product in glycolysis?

A: Pyruvic acid (two molecules/glucose molecule)

Q: What is the net yield (number of molecules) of ATP per glucose molecule in glycolysis?

A: Two. Four molecules of ATP are generated, but two are expended.

Q: In fermentation, pyruvic acid is converted into fermentation end products. How many additional molecules of ATP are generated per molecule of pyruvic acid by this process?

A: Zero. No ATP or other high-energy compounds such as $NADPH_2$ are produced by this process.

Q: With many bacteria, mutant strains can appear that are not able to synthesize an essential nutrient. These bacteria, therefore, must be supplied with this nutrient. What term is used to describe these mutants?

A: Auxotroph. The parental strain that can synthesize this nutrient is called a prototroph. Do not confuse auxotroph with autotroph. An autotroph is an organism that uses carbon dioxide as its primary source of carbon.

Q: What major family of potentially pathogenic gram-negative bacteria is oxidase negative?

A: Bacteria in the family Enterobacteriaceae (*E. coli*, *Salmonella*, *Shigella*, *Klebsiella*, etc.) are oxidase negative. Although there are exceptions, most other clinically significant bacteria are oxidase positive.

Q: Name some urease positive bacteria.

A: *Proteus*, *Klebsiella*, *Helicobacter*, and *Ureaplasma*

Q: How is *Proteus* easily differentiated from *Klebsiella*?

A: *Proteus* is motile (it swarms across the surface of some types of agar media such as blood agar), whereas *Klebsiella* is nonmotile.

Q: What complications can occur in cystitis due to urease-positive bacteria?

A: Urinary calculi (stone) formation. The urease splits the urea into ammonia and carbon dioxide. The ammonia increases the pH, resulting in precipitation of divalent cations such as magnesium and calcium.

Q: Most bacterial capsules are composed of polysaccharide. What is one major exception?

A: *Bacillus anthracis*, which has a capsule composed of D-glutamic acid

Q: What common gram-negative aerobic bacillus produces blue-green coloration on agar plates (and sometimes at infection sites)?

A: *Pseudomonas aeruginosa*. *P. aeruginosa* strains produce two water-soluble pigments, pyoverdin (green and fluorescent) and pyocyanin (blue). The combination appears blue-green. Pus from these infections may be similarly colored. (Trivia: Pyoverdin is a siderophore [iron-binding protein]; therefore it is a virulence factor for *P. aeruginosa*.)

Q: What gram-negative bacteria are sometimes referred to as the "black-pigmented" anaerobes?

A: *Prevotella* and *Porphyromonas*. The colonies sometimes turn black after prolonged incubation on blood agar plates.

Q: How does the Centers for Disease Control and Prevention define Category A bioterrorism agents?

A: Category A bioterrorism agents are high-priority agents that include organisms that pose a risk to national security because they can be easily disseminated or transmitted from person to person, result in high mortality, have the potential for major public health impact, might cause public panic and social disruption, and require special action for public health preparedness.

Q: List the Category A bioterrorism agents.

A: *Bacillus anthracis*, *Yersinia pestis*, *Francisella tularensis*, *Clostridium botulinum* toxin, variola virus (smallpox), and the viral hemorrhagic fever viruses

Q: What infections do *Y. pestis* and *F. tularensis* cause?

A: Plague and tularemia, respectively

Q: List some viral hemorrhagic fever viruses.

A: Ebola, Lassa, Marburg, and Machupo, to name a few

Q: Name the opportunistic gram-negative bacillus that forms red colonies.

A: *Serratia*, especially *S. marcescens*, are actually quite common nosocomial pathogens.

Q: What is meant by selective and differential media?

A: Selective media allow the growth of some bacteria and inhibit the growth of others. Differential media allow differentiation between the bacteria that grow on a particular medium. Most selective media are also differential. For example, MacConkey agar contains bile salts and crystal violet that are inhibitory for most gram-positive and some gram-negative bacteria; thus it selects for certain gram-negative bacteria. Of the gram-negative bacteria that grow, such as *E. coli*, *Salmonella*, and *Shigella*, lactose fermenters (red colonies) can be differentiated from lactose nonfermenters (colorless colonies).

Q: What component of the lipopolysaccharide is toxic?

A: Lipid A

Q: Which bacterium demonstrates a characteristic "tumbling" (head-over-heels) motility when viewed in a wet mount such as dark-field or phase-contrast microscopy?

A: *Listeria monocytogenes*

Q: A gram-negative rod was isolated from a burn wound. Is this organism most likely to be motile or nonmotile?

A: Motile. Most bacteria, especially bacilli, are motile. The major exceptions for bacilli are *Shigella*, *Yersinia*, and *Klebsiella*. The first letters of these three spell SYK (pronounced "sick"). They are too "SYK" to move. You can also assume that all cocci (gram positive and negative) are nonmotile.

Q: List at least four obligately aerobic bacteria.

A: *Pseudomonas*, *Mycobacterium*, *Bacillus*, and *Corynebacterium* must be on your list. Others include *Legionella*, *Nocardia*, *Bordetella*, and *Brucella*. Most standardized exams do not expect you to know these last four.

Q: List at least three obligately anaerobic bacteria.

A: *Clostridium*, *Bacteroides*, and *Actinomyces* are most important. Others include *Peptostreptococcus* (often referred to as "anaerobic strep"), *Propionibacterium*, *Prevotella*, *Porphyromonas*, and *Fusobacterium*.

Q: Name two microaerophilic bacteria.

A: *Campylobacter* and *Helicobacter* are the most important. *Borrelia* is also microaerophilic, but you do not need to know that.

Q: Define "facultative anaerobe" and list several.

A: Facultative anaerobes can grow either in the presence or in the absence of oxygen. They generally grow best with oxygen. Many pathogenic bacteria are facultative anaerobes, including staphylococci, *Vibrio*, and all the bacteria in the family Enterobacteriaceae (*E. coli*, *Salmonella*, *Shigella*, etc.).

Q: What are aerotolerant anaerobes (also called oxygen indifferent)?

A: Bacteria that are true anaerobes but can grow, albeit slowly, in the presence of oxygen. The streptococci are examples. They do not use oxygen as the terminal electron acceptor (as do the facultative anaerobes), even if oxygen is present. Most standardized exams do not expect you to know the aerotolerant anaerobes.

Q: Most bacteria have cell walls. Which do not?

A: Mycoplasmas; therefore cell wall active antimicrobials are ineffective.

Q: Without looking at a previous answer, list at least three bacteria that are not culturable on laboratory media.

A: *Treponema pallidum*, *Chlamydia*, *Chlamydophila*, and all the rickettsia. Note: *Chlamydophila psittaci* and *Chlamydophila pneumoniae* were previously classified as species of *Chlamydia*.

Q: Many bacteria are nutritionally fastidious and must be given additional growth factors for cultivation on laboratory media. Name two.

A: *Haemophilus influenzae*, *Legionella*, and *Neisseria gonorrhoeae* are three important ones you should know.

Q: With respect to the answer given for the previous question, what are the growth requirements for each?

A: *H. influenzae*: X (heme) and V (NAD) factors (X and V are letters, not Roman numerals); *Legionella*: L-cysteine and ferric ion; *N. gonorrhoeae*: you do not need to know the specific growth requirements. Just remember that it can grow on chocolate or Thayer-Martin agar (which is a chocolate agar supplemented with antimicrobials to inhibit normal flora contaminants).

Q: What is chocolate agar?

A: It is a type of blood agar plate. Normally, sheep (or other) blood is added to agar that has been melted and cooled to about 45°C, resulting in a red blood agar plate. Chocolate agar is similarly prepared, but the blood is added while the agar is about 80°C. The higher temperature results in lysis of the red blood cells (releasing intracellular nutrients) and causing the medium to look brown (like chocolate). Chocolate agar is useful for growing bacteria that are nonhemolytic.

Q: List some pleomorphic bacteria.

A: *Corynebacterium*, *Listeria*, *Legionella*, and some anaerobes (both gram positive and negative). Examination questions concerning these organisms often describe the isolation of pleomorphic rods. More information will always be given to lead you to the correct answer.

Q: Most pathogenic bacteria grow best from 35 to 37°C. Name one that grows above 40°C.

A: As an aid in identification, both *Campylobacter* and *Pseudomonas aeruginosa* can be grown at 42°C and enterococci can be grown at 45°C.

Q: Name a bacterium that can grow, although slowly, at refrigerator temperatures.

A: *Listeria monocytogenes* infections are frequently associated with dairy foods. This organism can grow to high numbers in refrigerated foods that have become contaminated. In the laboratory, this method of "cold enrichment" has been used to help isolate the organism.

Q: What role does IgA protease secretion play in bacterial pathogenesis?

A: It aids in the colonization of mucous membranes by some bacteria. Several species of bacteria secrete IgA1 protease that inactivates secretory IgA1 (but not IgA2) at mucosal surfaces, thus interfering with the antibacterial effect of the antibody.

Q: Name some organisms that secrete IgA protease.

A: Four of the most important are *Neisseria gonorrhoeae*, *N. meningitidis*, *Haemophilus influenzae*, and *Streptococcus pneumoniae*.

Q: Define exotoxin.

A: Exotoxins are toxic peptides or proteins secreted by bacteria. Because of the structure of the gram-negative cell wall, the secretion of proteins is more complex than in gram-positive bacteria. Gram-negative bacteria have various secretion mechanisms, called secretion systems, that allow the exotoxins (and other bacterial proteins) to be secreted across the outer membrane. There are at least eight bacterial secretion systems to date. One classification scheme includes six of these systems. These six secretion systems are designated as types I to VI (Roman numerals). Types III and IV are particularly important in bacterial virulence.

Q: What is an enterotoxin?

A: An exotoxin that has its major effect in the intestinal tract. "Enteric" refers to the intestines.

Q: Name some enterotoxin-producing bacteria.

A: *Staphylococcus aureus*, *Shigella*, *Escherichia coli*, *Clostridium perfringens*, *Clostridium difficile*, *Vibrio cholerae*, *Bacillus cereus*, and *Yersinia enterocolitica*. This is not a complete list, but you do not need to know any others at this time.

Q: What is a superantigen?

A: It is a protein, usually secreted by bacteria, that can nonspecifically activate a large number of T cells (primarily CD4+ T cells). Superantigens bind to the outside (rather than the groove; no antigen processing is required) of the Vβ region of the T-cell receptor and the MHC II molecule on antigen-presenting cells. This results in the activation of 2–20% of all CD4+ T cells with the release of cytokines (IL-2 from T cells and IL-1 and tumor necrosis factor from macrophages). In its most severe manifestation, superantigen activation of T cells can lead to shock and death.

Q: What are some bacterial diseases that are due to superantigens?

A: Toxic shock syndrome and scalded skin syndrome due to *Staphylococcus aureus*, and streptococcal toxic shock syndrome and rheumatic fever due to *Streptococcus pyogenes*. There are others, and some suspected viral diseases, but these are the most likely to be on exams.

Q: Describe the structure and function of the type III secretion system of pathogenic gram-negative bacteria.

A: The type III secretion system resembles a syringe, with the "needle" projecting from the outer surface of the outer membrane. The "barrel" of the syringe spans the cytoplasmic membrane, periplasmic space, and the outer membrane. The type III secretion system allows pathogenic gram-negative bacteria to inject virulence proteins such as exotoxins directly into the cytoplasm of susceptible cells (e.g., intestinal mucosal cells). The type III secretion system is found in many gram-negative pathogens, including *E. coli*, *Yersinia*, *Salmonella*, *Shigella*, *Helicobacter*, and *Bordetella*.

Q: What kind of nutrient transport into the cell requires the expenditure of energy?

A: Active transport

Q: The transport of nutrients across the cytoplasmic membrane frequently requires the expenditure of energy. What kind of transport relies on protein carrier molecules in the membrane but does not require energy?

A: Facilitated diffusion

Q: What is the function of the binding proteins that are found in the periplasmic space of gram-negative bacteria?

A: They bind nutrients that diffuse through the outer membrane and then present these nutrients to transport proteins of the cytoplasmic membrane for final transport into the cytoplasm.

Q: In what growth phase are bacteria most susceptible to cell wall active antimicrobials?

A: Log phase. This is the phase of growth in which the bacteria are synthesizing peptidoglycan at the most rapid rate. If bacteria are not synthesizing peptidoglycan, there is nothing to inhibit. This is also true for antimicrobials, as well as physical agents, such as ultraviolet light, that inhibit protein or nucleic acid synthesis.

Q: What enzymes are lacking in many anaerobic bacteria and therefore account in part for the toxic effect of oxygen?

A: Superoxide dismutase and catalase. It should be noted that some anaerobes, such as *Bacteroides fragilis*, possess both of these enzymes and can survive for significant periods of time in the presence of oxygen. Although they survive in oxygen, they will not grow until the redox potential reaches a sufficiently low level. Low redox potentials exist in anaerobic environments, especially if there are reducing agents present.

Q: An acid-fast stain was performed on an unknown organism in a student microbiology lab. After viewing the slide, only blue cells were seen. What do you conclude?

A: No acid-fast bacteria are present. Acid-fast bacteria are red (or pink).

Q: Bacteria without an H antigen would not be able to do what?

A: Swim. The H antigens are the flagella.

Q: A specific antiserum (antibody) was prepared against a structural component of a particular species of pathogenic bacteria. When this antiserum was mixed with a suspension of the bacteria, there was a positive quellung reaction. Assuming this structural component is a virulence factor, what is its most important function?

A: Resistance to phagocytosis. The structure is the capsule. The quellung test utilizes antisera prepared against the polysaccharide capsule. In the presence of specific antisera, the capsule appears to swell. This question requires you to know what the quellung reaction is, that it involves the bacterial capsule, and that the capsule is antiphagocytic. Knowledge of the quellung reaction is not high yield.

Q: What enzyme can be used to make protoplasts?

A: Lysozyme. Protoplasts, derived from gram-positive bacteria, have no peptidoglycan.

Q: A recent immigrant from Kazan, Russia presented to the emergency room in obvious respiratory distress. A throat swab was obtained, streaked onto various kinds of bacteriological media, and incubated. One of the bacterial colonies was stained with methylene blue and viewed microscopically. Small red intracellular granules were observed. Based on this information, this patient may have what disease?

A: Diphtheria. The granules seen were metachromatic granules (volutin).

Q: A bacterium isolated from a soil sample was found able to form spores, but only if grown under anaerobic conditions. What is the most likely identification of this organism?

A: *Clostridium*. The only bacteria (of clinical significance) that can sporulate are members of the genera *Bacillus* (aerobe) and *Clostridium* (anaerobe).

Q: In a laboratory experiment, a mutant of *E. coli* was no longer able to obtain iron. Why?

A: It no longer synthesized siderophores. Siderophores are small peptides secreted by many bacteria that can chelate iron. Not all bacteria synthesize siderophores. For example, *Listeria* uses exogenously available ferric siderophores made by other organisms.

Q: A student made a Gram stain of *E. coli* in the lab. When she looked at it under the microscope, it appeared red. What, if anything, did she do wrong?

A: She did nothing wrong; *E. coli* is supposed to look red (usually pink).

Q: A bacterium that was isolated from a burn wound was able to grow by both fermentation and by aerobic respiration. How can this organism best be classified?

A: Facultative anaerobe

Q: A mesophile would be expected to grow best under what conditions?

A: Moderate temperatures. Although most bacteria can grow to some extent over a wide range of temperatures, almost all have a narrow optimal range. Mesophiles, which essentially consist of all human pathogens (there are very few exceptions), grow best at temperatures from about room temperature (25°C) to about body temperature (37°C).

Q: What is quorum sensing?

A: The short answer: cell-to-cell communication. More technical: cell density dependent gene expression. Because of its relationship to infectious processes, there is a good chance you will see this concept on a standardized exam. Many bacteria and eukaryotic microbes appear to produce small signaling molecules called autoinducers. When the cell population reaches a critical size, the concentration of autoinducers reaches a threshold level that can be "sensed" by all the organisms in the population. At this time the organisms all do something at the same time. This "something" is usually favorable for the population but would be a waste of energy if each organism did this "something" when the population was small. As an example, elastase is a virulence factor of *Pseudomonas aeruginosa* and is regulated by a quorum-sensing mechanism. Elastase production by a few cells in an infection would not be very useful and would probably be a waste of energy. However, elastase production by a large number of cells at an infectious site would be much more effective for the entire population of bacteria to break down elastase to use as a nutrient source. Other activities besides enzyme production that have been found in some bacteria to be controlled by quorum sensing include motility, biofilm formation (such as on the surface of catheters), and sporulation.

Q: What is a biofilm?

A: Biofilm consists of many organisms, bacteria and/or other organisms, that form a film on solid surfaces, usually in the presence of fluid flow. Examples of fluid flow in the human body are blood, saliva, and mucus secretion. The organisms can attach to surfaces and to each other by a self-secreted glycocalyx, capsular material, pili, extracellular DNA, or other microbial products/structures. Bacteria can also use host products, such as remnants of dead white blood cells, to form biofilms. Biofilms can consist of single or multiple species and typically are multiple layers thick.

Q: Why are biofilms important in infectious diseases?

A: After initial colonization/infection, microbes (bacteria, yeasts, etc.) commonly begin to form mats or biofilms at the site of infection, or elsewhere if they travel via the blood/lymphatics to other areas of the body. Organisms in biofilms are more protected from the immune system (e.g., antibodies and macrophages) than if they were individual cells circulating in the body (e.g., blood). For example, it is relatively easy for a macrophage to phagocytize a single floating cell, but it would be more difficult to phagocytize many cells that are tenaciously attached in multiple layers to a solid surface. Biofilms contribute to antibiotic resistance during therapy because it is more difficult for antimicrobial agents to diffuse through a multilayered biofilm than to reach a single floating cell. In addition, many of the bacteria in a biofilm are in stationary phase, resulting in tolerance, especially to bactericidal agents. Biofilms can also be involved in persistent infections, because the bacteria in biofilms are difficult to eliminate.

Q: List some examples of biofilms in infectious diseases.

A: You can name almost anything and be correct. Some are just hard to visualize. An obvious example is the dental plaque on the surfaces of teeth that consists of multiple species of bacteria that lead to tooth decay, gingivitis, and periodontitis. Biofilm formation in the respiratory tract by *Pseudomonas aeruginosa* in cystic fibrosis patients is another example. Microbial biofilms also form on catheters and other artificial implants. Endocarditis involves biofilm formation of the endocardium or heart valves.

Q: In what growth phase are bacteria most susceptible to penicillin?

A: Log phase. In most cases, it is most effective to kill bacteria with antibiotics (and by other methods such as with ultraviolet light) if the bacteria are actively metabolizing. With penicillin, for example, if the bacteria are not synthesizing peptidoglycan, there is nothing to inhibit. Penicillin inhibits peptidoglycan synthesis; it does not degrade it.

Q: Bacteria are able to "extract" the chemical energy that is stored in nutrients. In which one of the following metabolic steps is *no* energy in the form of ATP or NADH$_2$ produced: glycolysis (glucose to pyruvic acid), fermentation (pyruvic acid to fermentation end products), TCA cycle, aerobic respiration, or anaerobic respiration?

A: No energy in the form of ATP or NADH$_2$ is produced in the fermentation pathways that convert pyruvic acid to fermentation end products.

Q: It is theoretically possible to grow an obligate aerobe under anaerobic conditions? How?

A: Yes. Grow it in the presence of an alternative terminal electron acceptor such as nitrate. Under these conditions, it would be growing by anaerobic respiration, not fermentation. This is the same pathway that allows *E. coli* and related uropathogens to convert nitrate to nitrite. Not all bacteria are capable of reducing nitrate to nitrite; however, because about 90% of urinary tract infections are due to *E. coli* and related bacteria, the nitrate reduction test is useful in detecting urinary tract infections.

Q: What causes thymine dimer formation in bacteria?

A: Ultraviolet light. The bonds between adjacent thymines in the dimer are covalent bonds. Thymine dimers in DNA result in massive mutations that lead to cell death.

Q: What chemical agent kills bacteria by specific irreversible binding to tyrosine, resulting in inactivation of proteins such as essential enzymes?

A: Iodine

Q: By what other mechanism does iodine kill?

A: It also kills by oxidation of proteins. Oxidation of sulfhydryl groups results in the formation of disulfide bonds.

Q: What is chlorhexidine?

A: Marketed as chlorhexidine gluconate, it is an antiseptic commonly used to treat gingivitis. Oral decontamination may reduce the risk for ventilator-associated pneumonia. It is also used as a surgical scrub and presurgical skin disinfection of the patient. Chlorhexidine causes disruption of bacterial membranes.

Q: Name a gaseous chemical that can be used for sterilization.

A: Ethylene oxide is usually the one most think of first, but beta-propiolactone is another.

Q: From the previous question, what is the mechanism of action?

A: These are alkylating agents that damage proteins and nucleic acids. Specifically, they alkylate –OH and –NH$_2$ groups. Formaldehyde and glutaraldehyde are also alkylating agents. Alkylating agents are sporicidal.

Q: In what form is chlorine used for killing microbes?

A: Aqueous form. In water, chlorine forms hypochlorous acid, which is the active form. Sodium hypochlorite (bleach) is similar. 1:10 Dilution. Use 50:50 For HIV

Q: What is the mode of action for the killing effect of chlorine?

A: Chlorine (hypochlorous acid) causes the oxidation of sulfhydryl groups resulting in disulfide bonds in proteins. The result is the inactivation of essential proteins in microbes.

Q: What minimum pore size in a filter is most commonly used to ensure that a fluid is bacteria free?

A: 0.22 μm

Q: Can any bacteria pass through the above filter?

A: Yes, mycoplasmas. They are not common contaminants, so we usually don't worry about them when filtering liquid solutions. To retain mycoplasma, a 0.1-μm filter is recommended.

Trivia: The largest known bacterium is *Thiomargarita namibiensis*, "sulfur pearl of Namibia." It has a volume about 3 million times that of *E. coli* and can be seen with the naked eye (it is almost 1 mm in diameter). It was discovered in sea floor sediments off the coast of Namibia. Cells contain reflective white globules of sulfur, hence the derivation of its name. The name derivation of "pearl" comes from the Latin word *margarita*, meaning "child of the sea."

Q: What are the minimum autoclave parameters that are most commonly used to ensure sterilization?

A: 121°C, 15 lb/in^2, 15 minutes, but the time may be longer for bulky materials.

Pressure of ↑ temp ↓

Q: In autoclaving, what is the purpose of 15 lb/in²?

A: To ensure that the temperature of the steam reaches 121°C. A lower pressure would decrease the achievable temperature, requiring a longer time for sterilization.

Q: What is an iodophor?

A: A combination of iodine and a detergent (such as a quaternary ammonium compound). Wescodyne® is an example.

Q: What is the purpose of pasteurization?

A: The parameters for pasteurization are designed to kill pathogenic bacteria, such as *Brucella*, *Mycobacterium bovis*, and so on, that can be found in dairy products, juices, or other consumable liquids. It does not sterilize. *Brucella* is not a significant pathogen in the United States but is significant in some developing countries.

Q: A relatively new method for sterilizing foods, such as hamburger and lettuce, utilizes high-energy electrons. This is an example of an ionizing radiation. Name two other forms of ionizing radiation and describe the mechanism of killing.

A: Two others are x-rays and gamma rays. They kill cells by generating hydroxyl radicals that damage intracellular components (primarily nucleic acids).

Q: What are quaternary ammonium compounds and how do they kill?

A: They are cationic detergents (long hydrophobic hydrocarbon tail that has a positive charge at the other end). They kill by disrupting membranes. They are commonly used in many over-the-counter (OTC) preparations (mouthwashes, skin disinfectants, etc.).

Q: What antibiotic has a similar mode of action as the quaternary ammonium compounds?

A: Polymyxin. Its structure resembles a quaternary ammonium compound. Daptomycin has some structural similarities, but as described later, has a different mode of action.

Q: What is the mode of action of phenolic compounds? List several examples.

A: They cause protein denaturation in membranes. They are frequently used in disinfectants and antiseptics. Hexylresorcinol and hexachlorophene are examples.

Q: What concentration of alcohol has the best antimicrobial effect?

A: Seventy to 95%. One would think that 100% alcohol would kill best, but water seems to augment the killing effect. 70% isopropanol is most common.

Q: Bismuth is used in the treatment of what bacterial infection?

A: Infection due to *Helicobacter pylori*. The most important thing to remember about treatment is that a single agent is never used. The current therapy consists of a combination of a proton pump inhibitor + clarithromycin + either metronidazole or amoxicillin. An alternative is bismuth subsalicylate + metronidazole + tetracycline + either a proton pump inhibitor or H_2 blocker.

Q: What is the mechanism of killing by hydrogen peroxide?

A: It is another oxidizing agent (oxidation of sulfhydryl groups to form disulfide bonds in proteins).

Q: How do heavy metals such as mercury or silver kill bacteria?

A: They bind to sulfhydryl groups of proteins, thus interfering with normal functions.

Q: What are some physical methods that can be used for sterilization (i.e., sporicidal, virucidal, fungicidal, bactericidal, etc.)?

A: Autoclaving and ionizing radiation are the two you should know. Both can fail to sterilize if improperly used (e.g., time too short).

Q: What are some chemical methods that can be used for sterilization (i.e., sporicidal, virucidal, fungicidal, bactericidal, etc.)?

A: Chlorine and iodine compounds, as well as alkylating agents (formaldehyde, glutaraldehyde, ethylene oxide, and beta-propiolactone). As with the physical methods, these can fail to sterilize if improperly used. Sterilization is time and concentration dependent and may be affected by extraneous organic material.

Q: Peptidoglycan structure consists in part of tetrapeptide chains that often are cross-linked to each other. Name the enzyme that catalyzes the cross-linking reaction.

A: Transpeptidase (also called penicillin binding protein)

2 | Genetics

Q: What is transcription?

A: Transcription is the process of making a messenger RNA (mRNA) from the complimentary chromosomal DNA.

Q: What is translation?

A: Translation is the process of making a protein from mRNA.

Q: In general terms, what is a bacterial operon?

A: An operon is a group of genes that code for at least one protein (usually more) and includes genes that regulate the transcription of the protein encoding region. All the genes in an operon are cotranscribed in a single strand of mRNA. About half of all protein-coding genes of prokaryotes are located in operons. Genes that are placed together in operons can vary widely across different species of microbes. That is, genes that are in the same operon in one bacterium are often found in different operons in other bacteria.

Q: Describe the structure and function of the genes that compose a typical operon, specifically the promoter, operator, cistron, and terminator.

A: **Promoter:** This is the DNA region within the operon to which RNA polymerase binds. After binding, mRNA is transcribed. **Operator:** This is a section of DNA, near the promoter, to which a repressor molecule can bind. If the repressor molecule binds, mRNA is not transcribed. An inducer molecule, if present, can bind to the repressor molecule, resulting in inactivation of the repressor molecule. This inactivation of the repressor molecule allows for mRNA transcription. **Cistron:** This is the section (gene) of the operon that codes for a protein, such as an enzyme. The mRNA being transcribed is subsequently translated into a protein product. Frequently, several genes are present; therefore the term *polycistronic* is used to describe this section of the operon. Because products (one or more proteins) are being transcribed, these genes are also referred to as being structural genes (the enzyme is the "structure"). **Terminator:** This is a short section of DNA after the cistrons that is a signal to stop transcribing the DNA into mRNA. If this termination signal is missing, transcription of the next genes along the DNA strand might occur.

Note: There are variations on this model, especially with respect to regulation of transcription (positive and negative control, apoinducers, corepressors, etc.), but it is unlikely you need to know that level of detail.

Q: What is transformation?

A: It is the uptake of free naked DNA from the environment by bacteria. Uptake can be prevented by the addition of DNase, which degrades the DNA before uptake. Uptake cannot be prevented if a filter separates the DNA from the bacteria unless the filter pore size is sufficiently small that the DNA fragments will not pass. DNA fragments can pass through a filter pore size of 0.45 μm. This pore size also allows the passage of bacteriophages, but not the bacterial cells.

Q: What name is given to small extrachromosomal genetic elements found in most bacteria?

A: Plasmids. Many plasmids carry genes that code for such characteristics as antibiotic resistance or virulence factors (e.g., toxins). Most plasmids are dispensable and are not required for growth or survival of the bacteria.

Q: What is the name for the exchange of genetic information between two bacterial cells that requires cell-to-cell contact?

A: Conjugation

Q: In bacterial conjugation, what designation is given to the donor and recipient cells?

A: Donor cells are males and are frequently called F^+. In some cases, the cells are called Hfr. The recipient cells are called females, or F^-.

Q: What does the "F" refer to in the previous question?

A: The F refers to a plasmid called a fertility factor (F factor, F plasmid). It carries the genes necessary to code for its own transfer (such as coding for sex pili) by conjugation to another cell. F factors do not necessarily code for antibiotic resistance. If they do, they are called resistance transfer factors.

Q: What are Hfr cells?

A: Hfr stands for high frequency of recombination. Hfr cells are male cells in which the F plasmid has integrated into the bacterial chromosome. As a functional male, it can participate in conjugation with an F⁻ cell. In conjugation between an Hfr and an F⁻ cell, the bacterial chromosome is transferred before the integrated plasmid is transferred. Because conjugation frequently occurs between donors and recipients of the same species, the chromosomal DNA that is being transferred has high homology with the recipient cell. Because of the high homology, the DNA from the donor can recombine into the recipient chromosome at a high frequency.

Q: What is an F′ (F prime) cell?

A: An F′ cell is a male cell in which the F plasmid integrates into the bacterial chromosome. This integration is reversible (i.e., the integrated F plasmid can come back out of the chromosome into the cytoplasm). This F plasmid now carries a piece of the chromosome, possibly several chromosomal genes, and the cell (and the plasmid) is now called F′. An F′ cell can function as a male in conjugation. During conjugation, the attached chromosomal DNA can be transferred into the F⁻ cell.

Q: Give an example of plasmid encoded exotoxin production.

A: Some examples are the exotoxins of *Clostridium tetani* and *Bacillus anthracis*, the heat-labile and heat-stable toxins of *Escherichia coli*, and exfoliatin of *Staphylococcus aureus* (toxin in staphylococcal scalded skin syndrome).

Q: In a genetics experiment, it was found that DNA transfer from one strain of *E. coli* to another strain would occur in a solution even if the two strains were separated by a filter (0.45 μm pore size) and DNase was added to the solution. What is the most likely mechanism of this genetic transfer?

A: Transduction. In transduction, the DNA from one strain of bacteria is transferred to another strain via a bacteriophage vector. Phages are small enough to pass through most filters. Any DNA inside the phages is protected from the DNase by the viral capsid.

Q: Define lysogeny.

A: Lysogeny is the stable carriage of bacteriophage DNA, generally in the form of a prophage, by bacteria.

Q: Define a prophage.

A: A prophage is bacteriophage DNA that becomes integrated into the host bacterial chromosome in a stable form.

Q: Define lysogenic conversion.

A: Lysogenic conversion is the expression of a new phenotypic characteristic by a bacterial cell due to infection of the cell with a lysogenic bacteriophage.

Q: List some examples of exotoxin production by bacteria due to lysogenic conversion.

A: Examples are the exotoxins produced by *Corynebacterium diphtheriae*, *Clostridium botulinum*, and *Vibrio cholerae*; the pyrogenic (erythrogenic) toxins of *Streptococcus pyogenes*; the toxic shock syndrome toxin and enterotoxins of *Staphylococcus aureus*; and the shiga-like toxin of *E. coli* O157:H7.

Q: Describe the structure of a typical bacterial chromosome.

A: The typical bacterial chromosome consists of double-stranded DNA that is circular rather than linear, as is found in eukaryotic cells.

Q: Describe the "ploidy" with respect to the bacterial chromosome.

A: Bacteria are haploid, which means there is only one copy of the bacterial chromosome. At times during the replication cycle there may be more than one copy, but only one copy is required.

Q: What are the genetic elements called that can "hop" from one location on the chromosome or a plasmid to another location?

A: Both insertion sequences and transposons can do this. These are sometimes referred to as mobile genetic elements.

Q: What is transfection?

A: Transfection is a common technique used in molecular biology to transfer DNA or RNA into eukaryotic cells. It is similar to transformation in bacteria. To make things confusing, there is a process of transfection in bacteria using phage DNA, but this is advanced molecular biology that will not appear on standardized exams.

Q: In a genetics experiment, it was found that DNA transfer between two strains of *E. coli* could be prevented if the two bacterial strains were separated by a filter (0.45 µm pore size). What is this type of genetic transfer?

A: Conjugation. The filter prevents the two strains from making contact.

Q: In bacterial conjugation between an F$^+$ cell and an F$^-$ cell, what happens to the phenotype of the F$^+$ cell after conjugation is completed (i.e., does it remain F$^+$, is it converted to an F$^-$, does it die, etc.)?

A: Nothing; it remains as an F$^+$ cell. The F$^+$ cell acts as a male donor and only donates a copy of the F plasmid. The original F plasmid stays in the F$^+$ cell.

Q: In conjugation between an Hfr cell and its recipient cell, what *typically* happens to the *recipient* cell?

A: It remains as an F$^-$ cell. In this type of conjugation, the Hfr donor cell has the F plasmid integrated into the cell chromosome. The Hfr cell still functions as a male, but instead of donating a copy of the F plasmid, chromosomal DNA from the Hfr cell first enters the recipient F$^-$ cell. In nature, the conjugal bridge breaks between the Hfr cell and the F$^-$ cell before too much of the chromosomal DNA is transferred. Because the F plasmid usually is not transferred, the recipient remains an F$^-$ cell. If the two cells were allowed to remain together so that conjugation could be completed, then the last piece of DNA that enters the F$^-$ cell would be the F plasmid. Under these conditions (usually under controlled conditions in a laboratory), the recipient female could be converted to a male donor.

Q: In tracking down the specific physical location of a gene that codes for beta-lactamase production in a plasmid-free strain of *Staphylococcus aureus*, some contradictory results were obtained. In some clones, the gene appeared to be in one location on the chromosome, but in other clones, the gene was found in several additional locations on the chromosome. How can this best be explained?

A: The gene was part of a transposon and was able to "hop" to different locations.

Q: Review: How is the DNA transferred from one cell to another in transduction?

A: By bacteriophages

Q: It was observed during one set of transduction experiments that random genes from different locations on the donor DNA were transferred by the transducing bacteriophage to the transduced (recipient) bacterial cells. What name is given to this type of transduction?

A: Generalized transduction

Q: In another set of transduction experiments it was observed that only one or two genes, both located next to each other on the donor DNA, were transferred by the transducing bacteriophage to the transduced bacterial cells. What name is given to this type of transduction?

A: Specialized transduction

Q: What is the difference between homologous and nonhomologous recombination?

A: In homologous recombination, such as in bacteria, a section of DNA from a donor cell enters (by conjugation, etc.) another genetically related (homologous) cell and easily inserts itself into the host chromosome (because they are closely related and have similar nucleotide sequences, i.e., homologous sequences). In nonhomologous recombination, a section of DNA from a donor cell enters a genetically unrelated (nonhomologous) cell and inserts itself into the host chromosome (because they are not closely related, it is more difficult for the new nonhomologous nucleotide sequences to integrate).

Antimicrobial Agents

Q: What is meant by MRSA?

A: MRSA stands for methicillin-resistant *Staphylococcus aureus*. Methicillin was first used in 1959; the first MRSA strain was detected in 1961.

Q: What is the clinical significance of MRSA?

A: MRSA strains are resistant to methicillin. Methicillin is a beta-lactamase–resistant beta-lactam antibiotic. Beta-lactamase production is one of the major mechanisms of resistance of bacteria to the beta-lactams. The beta-lactamase inactivates beta-lactam antibiotics by hydrolyzing the beta-lactam ring. Methicillin is one of several beta-lactams that is resistant to beta-lactamase; therefore it, and related beta-lactams, can be used as therapy even for beta-lactamase–producing bacteria. However, if bacteria such as *S. aureus* become resistant to methicillin, then none of the beta-lactamase resistant drugs can be used. MRSA strains are becoming common. In the United States approximately 60% of *S. aureus* infections in intensive care units are MRSA. Community-acquired MRSA infections continue to rise.

Q: What is the mechanism of resistance of MRSA?

A: Resistance is related to a chromosomal gene that codes for an abnormal penicillin-binding protein (transpeptidase). Transpeptidases catalyze the last step in peptidoglycan synthesis, which is the cross-linking of adjacent amino acid side chains. Bacteria can have several transpeptidases (some strains of *Escherichia coli* have four), each with a modified amino acid sequence different from the others. They are the targets for all beta-lactam antibiotics. In susceptible bacteria, the beta-lactams bind to, and therefore inactivate, the transpeptidase, thus stopping cell wall synthesis. In MRSA, one of the transpeptidases has a decreased affinity for binding beta-lactam antibiotics, resulting in resistance not only to methicillin but to all beta-lactams, including penicillins, cephalosporins, cephamycins, and carbapenems. In MRSA strains, resistance to non–beta-lactam antibiotics (erythromycin, clindamycin, gentamicin, trimethoprim-sulfamethoxazole, ciprofloxacin, etc.) is common.

Q: What is the drug of choice for MRSA?

A: Vancomycin

Q: Describe daptomycin, its mode of action, and its clinical use.

A: It is a newly approved antibacterial agent (a lipopeptide) with a spectrum of activity limited to gram-positive organisms, including MRSA, vancomycin-resistant *S. aureus*, and vancomycin-resistant *Enterococcus*. It has been approved for the treatment of skin and soft tissue infections. Its mode of action is that is causes disruption of the bacterial membrane through the formation of transmembrane channels. These channels cause leakage of intracellular ions, leading to depolarization of the cellular membrane and inhibition of macromolecular synthesis.

Q: An environmental microbiologist collected a water sample from an air conditioning cooling tower. A Gram stain was performed and gram-positive bacteria were observed. Would this organism be expected to be susceptible or resistant to penicillin?

A: There is no way of telling the antibiotic susceptibility pattern in this case based on the Gram stain morphology alone. The organism would need to be identified and then an antibiotic susceptibility test performed.

Q: A 7-year-old child with impetigo visited her pediatrician with her father. Pus from one of the blister-like lesions was collected and Gram stained. The technologist saw many gram-positive cocci in irregular clusters. Would this organism be expected to be susceptible or resistant to dicloxacillin?

A: In this case you can guess that it is most likely susceptible to dicloxacillin. The child has impetigo that presents with blisters; thus the child has bullous impetigo. Bullous impetigo is commonly due to *Staphylococcus aureus*. The gram-stain morphology supports, but does not confirm, the identity of this organism as *S. aureus*. Empiric therapy for community-acquired *S. aureus* infections includes the penicillinase-resistant beta-lactams, such as dicloxacillin.

If this strain of *S. aureus* is methicillin resistant (MRSA; a 20–40% possibility, depending on the local community prevalence), then it would be resistant to dicloxacillin.

Q: What antibacterial is effective against both growing and nongrowing (stationary) cells?

A: Membrane-active agents such as polymyxin cause disruption of the cytoplasmic membrane and therefore do not require active metabolism by the cells. Daptomycin is also effective against cells in stationary phase. Most other antimicrobials require active metabolism. With most others, if the cells are not metabolically active, there is nothing to inhibit.

Q: What is clavulanic acid?

A: It is a beta-lactamase inhibitor. When used in combination with a beta-lactam antibiotic such as amoxicillin, it "converts" a beta-lactam resistant organism into a susceptible organism. The resistant organism must be resistant because of the secretion of beta-lactamase. The clavulanic acid (usually in the form of sodium clavulanate) must be able to bind irreversibly to the beta-lactamase.

Q: What are two other beta-lactamase inhibitors?

A: Sulbactam and tazobactam

Q: What are some antibiotics that inhibit protein synthesis?

A: Some of the most noteworthy are the aminoglycosides, tetracyclines, chloramphenicol, lincosamides (e.g., clindamycin), and the macrolides.

Q: Draw a graph showing the effect on the growth curve of bacteria if a bactericidal antibiotic was added at mid–log phase. Now show the effect if the antibiotic was bacteriostatic.

A: Figure 3-1 is an example of what the graph should resemble. The arrow shows when either a bactericidal or bacteriostatic antibiotic is added. There is always a short lag time after addition before the cells respond. The response time and the specific shape of the curves vary depending on the species/strain of organism, specific agent, concentration, temperature, and so on.

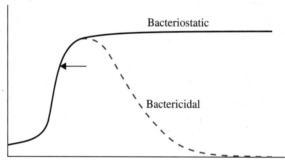

Figure 3-1

Q: Activation of autolytic enzymes leading to cell death by lysis occurs with what antimicrobials?

A: The beta-lactams. These agents not only inhibit peptidoglycan synthesis, which by themselves may not be bactericidal, but also cause an overproduction or other activation of normal autolytic enzymes in the bacteria. The increased level/activity of the autolytic enzymes contributes to cell death after the antibiotic inhibits further cell wall synthesis. These autolytic enzymes are normally synthesized at a low level and cause controlled lysis of the peptidoglycan. In this way, new peptidoglycan components can be added as the cell grows.

Q: What, in general, is the mode of action of penicillin?

A: It inhibits cell wall synthesis, specifically peptidoglycan.

Q: What enzymatic reaction is inhibited by penicillin?

A: Transpeptidation, which normally results in cross-linked peptidoglycan. The enzyme inhibited is called transpeptidase.

Q: What is the normal function of penicillin binding proteins in bacteria?

A: Transpeptidation, which results in covalent bond formation between tetrapeptide side chains of peptidoglycan. The term *penicillin binding protein* and transpeptidase are interchangeable.

Q: What is the mechanism of resistance of bacteria to penicillin?

A: Resistance to penicillin and other beta-lactam antibiotics is primarily due to beta-lactamase production by the bacteria. Other mechanisms of resistance include decreased permeability as well as modification of the penicillin binding proteins, which prevents binding to the penicillin binding protein. The latter mechanism of resistance is seen in MRSA.

Q: List some organisms for which the penicillins are commonly drugs of choice.
A: Generally, gram-positive cocci, including both aerobes and anaerobes; most other anaerobes (except *Bacteroides fragilis* and *Clostridium difficile*); and *Treponema pallidum* (syphilis). For *Staphylococcus aureus* the usage varies geographically, but a penicillinase-resistant penicillin is frequently a good choice for empiric therapy until susceptibility results are available.

Q: Why is penicillin *not* a drug of choice for *Bacteroides fragilis*?
A: *B. fragilis* constitutively (i.e., even in the absence of penicillin) produces a beta-lactamase, making it naturally resistant to both the penicillins and the cephalosporins.

Q: What term is used for agents, such as some antibiotics, that kill rather than inhibit bacteria?
A: Bactericidal. Examples of antibiotics are those that inhibit cell wall synthesis or cause damage to the cytoplasmic membrane. The aminoglycosides are also bactericidal, as are several other drugs not listed here.

Q: Why does penicillin not affect mammalian cells?
A: Mammalian cells do not synthesize peptidoglycan (murein), which is the target for all cell wall active antibiotics.

Q: Antimicrobial resistance can be transferred among bacteria by a number of different mechanisms. Name the three basic mechanisms.
A: Transformation, conjugation, and transduction

Q: Metronidazole is used for infections involving what kinds of organisms?
A: Infections involving anaerobic bacteria, such as *Bacteroides fragilis* (and related bacteria), *Clostridium tetani*, and *Clostridium difficile*; protozoan infections, especially trichomoniasis, giardiasis, and amebiasis; and *Helicobacter pylori* (in combination with other agents such as a proton pump inhibitor and clarithromycin). It is also a drug of choice for bacterial vaginosis.

Q: What is the mode of action of metronidazole?
A: For most standardized exams, you will not need to know this. For those of you who want to know, after entering cells by passive diffusion, a nitro group is reduced to a highly reactive cytotoxic nitro radical that primarily causes damage to the DNA. This reduction of the nitro group occurs at a low redox potential that is found in anaerobic bacteria and some parasites. Therefore obligate aerobes and facultative anaerobes are generally resistant because the intracellular redox potential is too high for the reduction to occur. In the answer to the previous question you will notice that *H. pylori*, a microaerophile, is susceptible to metronidazole. This is unusual because typically even the presence of a small amount of oxygen makes an organism resistant. As it turns out, *H. pylori* contains a novel oxidoreductase that can reduce metronidazole to its active form in the presence of oxygen.

Q: What is the mode of action of the aminoglycosides?
A: They inhibit protein synthesis. Specifically, they have several effects on protein synthesis, including blocking the initiation of protein synthesis, causing premature termination of translation (incomplete protein), and incorporation of the wrong amino acid (misreading).

Q: What is the intracellular target of the aminoglycosides?
A: They bind to the 30S subunit of the 70S ribosomes.

Q: Are aminoglycosides bactericidal or bacteriostatic?
A: Bactericidal. The binding to the 30S ribosome is irreversible.

Q: Name some aminoglycosides.
A: Amikacin, gentamicin, tobramycin, kanamycin, streptomycin, tobramycin, netilmicin, and neomycin

Q: In general, aminoglycosides are most commonly used for what bacteria?
A: Aerobic gram-negative bacteria. Anaerobes are resistant because entrance into bacteria is an oxygen-linked transport system. There are some gram-positive bacteria that are susceptible, but other antimicrobials are more frequently used. Streptomycin can be used in combination with other drugs for the treatment of tuberculosis.

Q: What is the mechanism of resistance of bacteria to gentamicin?

A: You need to know that gentamicin is an aminoglycoside because questions on standardized exams may ask about modes of action and resistance to the drug group (aminoglycosides) or to major drugs within the group (gentamicin, streptomycin, tobramycin, amikacin, netilmicin, etc.). There are three mechanisms of resistance for aminoglycosides: (1) most important of the three is enzymatic modification of the drug (phosphorylation, adenylation, or acetylation), (2) decreased uptake, and (3) an alteration of the ribosomal binding site.

Q: What kind of antimicrobials are the cephalosporins?

A: They are beta-lactams.

Q: What is meant by "generation" when referring to the cephalosporins?

A: There are four generations of the cephalosporins. The antibiotics in each generation have been chemically modified in such a way that their spectrum of activity has been modified, generally to cover more gram-negatives. In general, first-generation drugs are active against many gram-positives and a few gram-negatives, second-generation drugs are active against fewer gram-positives and more gram-negatives, third-generation drugs are active against even fewer gram-positives and more gram-negatives, and fourth-generation drugs are active against many gram-positives and gram-negatives, including some beta-lactamase producers.

Q: What is the mechanism of resistance of bacteria to ceftriaxone?

A: Ceftriaxone is a third-generation cephalosporin (you should know some of the major cephalosporins, but not necessarily in what generation they should be classified). Resistance to the cephalosporins and other beta-lactam antibiotics is primarily due to beta-lactamase production by the bacteria. Some gram-negative bacteria secrete enzymes known as "extended spectrum beta-lactamases" that are very active against some beta-lactamase–resistant cephalosporins such as ceftazidime and cefotaxime and aztreonam (a monobactam).

Q: What are the cephamycins?

A: The cephamycins are a variant of the cephalosporins with a methoxyl group on the beta-lactam ring, rendering them more resistant to penicillinase. In most cases, all you need to know is that they are beta-lactam antibiotics related to the cephalosporins. Cefoxitin and cefotetan are included in this group.

Q: What are the carbapenems?

A: As with the previous question, knowing that this is a beta-lactam antibiotic is probably enough. Examples include imipenem and meropenem.

Q: What is the postantibiotic effect?

A: The postantibiotic effect, or PAE, refers to a period of time after complete removal of an antibiotic during which there is no growth of the targeted bacterium. Most antimicrobial agents have a PAE on a wide variety of bacterial pathogens. The duration of the PAE is due primarily to the bacterial species, as well as the type and concentration of the antibiotic.

Q: How does the PAE of an antibiotic affect therapy?

A: The presence of a long PAE, such as with the aminoglycosides, allows for less frequent dosing. Antibiotics that do not have a substantial PAE or have a short PAE, such as with many beta-lactams, require frequent or continuous dosing.

Q: What is the mechanism of action of chloramphenicol?

A: It inhibits protein synthesis by binding to the bacterial 50S ribosomal subunit and inhibiting peptidyl transferase; thus there is no peptide bond between the new amino acid and the remaining portion of the growing peptide. You should be familiar with the chloramphenicol-associated toxicities, such as aplastic anemia. Pharmacology is not covered extensively in this book, but drug toxicities, interactions, and so on are common questions on standardized examinations. Be sure to review these.

Q: What is the mechanism of resistance of bacteria to chloramphenicol?

A: Inactivation of the drug due to acetylation. Resistant bacteria synthesize a plasmid-encoded acetyltransferase.

Q: For what infections should chloramphenicol be used?

A: For serious infections when other antimicrobials cannot be used. It is rarely the first drug of choice. That is really all you need to know about chloramphenicol usage from a microbiological perspective. There are some pharmacological issues about chloramphenicol that you should review (that is why it is reserved for serious infections only). For those of you who must know its usage, it is an alternative for typhoid fever, *H. influenzae,* meningococcal and pneumococcal meningitis, *Bacteroides*, *Clostridium perfringens*, tularemia, plague, psittacosis, and rickettsial infections, to name a few.

Q: **What kind of drug is ciprofloxacin and what is its mode of action?**
　A: It is a fluoroquinolone that inhibits DNA synthesis by binding to DNA gyrase (topoisomerase). Other fluoroquinolones are levofloxacin, ofloxacin, trovafloxacin, gatifloxacin, moxifloxacin, norfloxacin, lomefloxacin, and enoxacin. The fluoroquinolones are derivatives of the older quinolones (nalidixic acid) and have a broad spectrum of activity.

Q: **For what infections should the fluoroquinolones be used?**
　A: As mentioned in the previous answer, the fluoroquinolones have a broad spectrum of activity. Some uses include soft tissue infections due to *S. aureus*, including some due to MRSA, pneumococcal pneumonia, urinary tract infections, meningococcal meningitis, campylobacteriosis, typhoid fever (a drug of choice), shigellosis (drug of choice), and rickettsia.

Q: **What is the mechanism of resistance of bacteria to the fluoroquinolones?**
　A: Resistance is usually due to mutations in the genes that code for the DNA gyrase, leading to modifications that prevent binding of the drug, or mutations that result in active efflux (the antibiotic enters the cells, but it is actively pumped back out). As with most other drugs, resistant strains of normally susceptible strains are being reported with increasing frequency. It has been strongly suggested that this increase in resistance is partly due to the use of fluoroquinolones (some not used in human medicine) in the poultry industry. As a result, cross-resistance to fluoroquinolones used in human medicine has developed.

Q: **What antimicrobial agents act as metabolic analogs and competitively inhibit the synthesis of para-aminobenzoic acid?**
　A: The primary agents are the sulfonamides, but there are other drugs, such as para-aminosalicylic acid and dapsone (a sulfone), that have this mechanism. The sulfonamides, sulfones, and para-aminosalicylic acid are analogs of para-aminobenzoic acid.

Q: **Why do the sulfonamides have no effect on human cells?**
　A: Sulfonamides inhibit the synthesis of tetrahydrofolic acid, which participates in the synthesis of folic acid. Humans do not have this enzyme. Folic acid is ingested preformed in our diet.

Q: **A burn wound patient developed an infection due to *Pseudomonas aeruginosa*. What is the therapy?**
　A: Combination therapy is commonly used because of the high resistance rate for this organism. An aminoglycoside (such as tobramycin, gentamicin, or amikacin) and a beta-lactam (such as ticarcillin/clavulanate or piperacillin/tazobactam) are commonly used for most non–urinary tract *P. aeruginosa* infections.

Q: **What is the drug of choice for a *P. aeruginosa* urinary tract infection?**
　A: Ciprofloxacin, but it is not recommended for children or during pregnancy

Q: **What are extended-spectrum beta-lactamases (ESBLs)?**
　A: They are enzymes that mediate resistance to extended-spectrum (third-generation) cephalosporins (e.g., ceftazidime, cefotaxime, and ceftriaxone) and monobactams (e.g., aztreonam). They do not affect cephamycins (e.g., cefoxitin and cefotetan) or carbapenems (e.g., meropenem or imipenem). ESBLs are found in the family Enterobacteriaceae, especially the genera *Klebsiella* and *Escherichia* (*E. coli*). These enzymes are rare in other bacteria.

Q: **What is the treatment for tuberculosis?**
　A: Isoniazid, rifampin, pyrazinamide, and ethambutol for 6–9 months is a recommended primary drug combination. Second-line drugs include cycloserine, ethionamide, p-aminosalicylic acid, streptomycin, capreomycin, amikacin/kanamycin,* levofloxacin,* moxifloxacin,* and gatifloxacin* (* = not approved by the U.S. Food and Drug Administration [FDA] for tuberculosis).

Q: **What is extensively drug-resistant tuberculosis?**
　A: The World Health Organization (WHO) has defined extensively drug-resistant tuberculosis as the occurrence of tuberculosis in persons whose *M. tuberculosis* isolates are resistant to isoniazid and rifampin and have resistance to any fluoroquinolone and at least one of three injectable second-line drugs (i.e., amikacin, kanamycin, or capreomycin). Because the definition is so long, it is doubtful that you will need to memorize it for standardized exams. You should, however, be aware that resistant strains are becoming common in endemic areas, including areas where HIV infection is high. It has been observed that resistance to the fluoroquinolones and second-line injectable drugs has been associated with poor treatment outcomes.

Q: What is the primary reason that combination chemotherapy is used in the treatment of tuberculosis?

A: It is used to prevent the development of antimicrobial resistant strains.

Q: How does isoniazid work?

A: It inhibits mycolic acid synthesis in the cell walls of *Mycobacterium*. This effect is bactericidal for the cells since the mycolic acid is an essential structural component of the cell wall.

Q: What antimicrobials are most effective against *Bacteroides fragilis* and related bacteria?

A: The *B. fragilis* group is composed of anaerobes. All in the group are beta-lactamase positive but respond well to metronidazole (drug of choice). Clindamycin can also be used, but resistance is emerging. A beta-lactam in combination with a beta-lactamase inhibitor can also be effective.

Q: Vancomycin is a large molecule (molecular weight >1400) that is ineffective against gram-negative bacteria. Why?

A: The outer membrane porins of gram-negative bacteria do no allow vancomycin to diffuse through the outer membrane.

Q: What group of antibiotics can be used to make protoplasts?

A: Cell wall active antibiotics. Protoplasts are bacterial cells that have no peptidoglycan. By growing susceptible bacteria in the presence of cell wall active antibiotics, many species can form protoplasts. These must be kept in an isotonic solution to prevent cell lysis.

Q: What drugs should be used for prophylaxis for *Pneumocystis jiroveci* (formerly *carinii*) pneumonia (PCP)?

A: Trimethoprim-sulfamethoxazole is the drug of choice. Dapsone or dapsone + pyrimethamine + leucovorin are also effective. Pentamidine is inhaled in an aerosol form to prevent PCP. Pentamidine is also used intravenously to treat active PCP. Atovaquone is a drug used in people with mild or moderate cases of PCP who cannot take trimethoprim-sulfamethoxazole or pentamidine. Because of the name change from *carinii* to *jiroveci* (pronounced yee-roo-vet-zee), PCP now stands for *Pneumocystis* pneumonia. *Pneumocystis carinii* still exists, but it is found only in rats. Therapy can be discontinued for patients receiving highly active antiretroviral therapy with an increase in CD4+ count to >200 cells/mm^3 for at least three months.

Q: What is the drug of choice for *Staphylococcus aureus* if it produces penicillinase? If it does not produce penicillinase? If it is MRSA?

A: **Penicillinase-producing:** a penicillinase-resistant penicillin such as cloxacillin, dicloxacillin, nafcillin, or oxacillin. **Non-penicillinase-producing:** penicillin G or V. **MRSA:** Vancomycin should be the major drug learned, but it can be used in combination with gentamicin and/or rifampin.

Q: When is the use of itraconazole indicated?

A: Itraconazole is indicated for the treatment of fungal infections such as blastomycosis, coccidiomycosis, paracoccidiomycosis, histoplasmosis, aspergillosis, sporotrichosis, and onychomycosis (due to dermatophytes) in both immunocompromised and non-immunocompromised patients.

Q: What is the mechanism of action of itraconazole?

A: Itraconazole is an azole, as is ketoconazole. Both inhibit cytochrome P450-dependent 14-alpha-demethylase, which is involved in the synthesis of ergosterol. Ergosterol is a component of fungal cell membranes. The result is structural changes to the membrane with concomitant growth inhibition and interference of membrane function, including interference of membrane-bound enzymes.

Q: What is the mode of action of amphotericin B?

A: Amphotericin B is a polyene antimicrobial that binds to ergosterol that is found in the cell membranes of fungi, resulting in membrane damage and leakage, particularly leakage of potassium.

Q: When is the use of ketoconazole indicated?

A: Ketoconazole is indicated for the treatment of systemic fungal infections such as candidiasis, chronic mucocutaneous candidiasis, oral thrush, blastomycosis, coccidioidomycosis, histoplasmosis, chromomycosis, and paracoccidioidomycosis, as well as some severe cutaneous dermatophyte infections.

Q: What is the mechanism of action of ketoconazole?

A: It impairs the synthesis of ergosterol in the cell membranes by the same mechanism as itraconazole.

Q: When is the use of flucytosine indicated?

A: Flucytosine is used in combination with amphotericin B to treat serious fungal infections caused by *Candida* or *Cryptococcus*.

Q: What is the mode of action of flucytosine?

A: It inhibits protein and DNA synthesis. Specifically, it inhibits protein synthesis by replacing uracil with 5-fluorouracil in RNA and it inhibits DNA synthesis by inhibiting thymidylate synthetase (you do not need to know this for standardized exams).

Q: When is the use of nystatin indicated?

A: Nystatin is used for *Candida albicans* infections, such as oropharyngeal and vulvovaginal candidiasis. Currently, it is topical only, but a liposomal systemic form may become available for other fungal infections.

Q: What is the mechanism of action of nystatin?

A: Nystatin is another polyene antimicrobial (like amphotericin B) that acts by binding to ergosterol in the cell membrane of fungi, resulting in a change in membrane permeability, thereby causing leakage of intracellular components.

Q: What is the mode of action of griseofulvin?

A: Griseofulvin acts primarily by disrupting the mitotic spindle structures of fungal cells. It is used to treat fungal infections of the skin, hair, fingernails, and toenails. Because of toxicity issues, griseofulvin use is not common and has been replaced with less toxic drugs such as itraconazole and terbinafine.

Q: What drugs can be used for the treatment of genital herpes?

A: Genital herpes is due to herpes simplex viruses type 1 and type 2. Most genital herpes is caused by type 2. There is no cure for herpes, but antiviral medications can shorten and prevent outbreaks. Suppressive therapy for symptomatic herpes can reduce transmission to partners. Acyclovir is commonly prescribed, but others include famciclovir, valacyclovir, vidarabine (adenosine arabinoside), trifluridine, and idoxuridine (iododeoxyuridine). Penciclovir cream is useful for mucocutaneous infections (herpes labialis, i.e., cold sores). Docosanol cream is a long-chain alcohol recently approved for over-the-counter treatment for herpes labialis (cold sores).

Q: What is the mechanism of action of acyclovir?

A: It selectively inhibits the replication of herpes viruses. Acyclovir is phosphorylated by viral thymidine kinase. The phosphorylated drug inhibits the viral DNA polymerase. Acyclovir can also cause DNA chain termination after incorporation of the phosphorylated form into the viral DNA.

Q: Name the classes of drugs for HIV infection.

A: Nucleoside reverse transcriptase inhibitors (NRTIs), nucleotide reverse transcriptase inhibitors (NtRTIs), non-nucleoside reverse transcriptase inhibitors (NNRTIs), protease inhibitors (PIs), and fusion inhibitors

Q: Name three NRTIs.

A: The seven approved for use are abacavir, didanosine, emtricitabine, lamivudine, stavudine, zalcitabine (dideoxycytidine), and zidovudine (azidothymidine).

Q: How do NRTIs work?

A: NRTIs are nucleoside analogs that work in the early phase of viral replication. Within the infected cells, these agents are phosphorylated before binding to the viral reverse transcriptase and incorporating themselves into the viral DNA strand. NRTIs eventually result in DNA chain termination. NtRTIs work similarly, but do not need to be phosphorylated (they are already phosphorylated).

Q: Name some NNRTIs.

A: Nevirapine, efavirenz, and delavirdine are three approved for use.

Q: How do NNRTIs work?

A: NNRTIs inhibit HIV virus replication by directly binding to and inhibiting the reverse transcriptase enzyme. These agents inhibit only the replication of HIV-1.

Q: Name some PIs.

A: Amprenavir, atazanavir, darunavir, fosamprenavir, indinavir, lopinavir (available in the United States only in combination with ritonavir), nelfinavir, ritonavir, saquinavir, and tipranavir. Ritonavir is usually taken in addition with other PIs (atazanavir, darunavir, fosamprenavir, saquinavir, tipranavir).

Q: How do PIs work?

A: PIs block the protease enzyme in a late stage of the viral replication process, resulting in mature viral particles that are defective and unable to infect new cells.

Q: Name a fusion inhibitor.

A: There is only one, enfuvirtide.

Q: How does enfuvirtide work?

A: It binds to the transmembrane glycoprotein subunit (gp41) of the viral envelope, which prevents the conformational change (fusion of the viral and host cell membranes) required for entry of the virus into the cell.

Q: In general, what is the recommended antiretroviral treatment of HIV infection?

A: One PI and two NRTIs. The FDA has approved a once-a-day three-drug combination tablet for treatment of HIV-1. The tablet contains efavirenz (an NNRTI), emtricitabine (an NRTI), and tenofovir disoproxil fumarate (an NtRTI).

Q: When should resistance testing be done with respect to HIV therapy?

A: Before starting antiretroviral therapy

Q: How do you treat the first episode of genital herpes?

A: Acyclovir is a good first drug of choice. It is administered orally three or five times a day (depending on the dosage) for 7–10 days. Famciclovir or valacyclovir are also recommended drugs of first choice.

Q: What antivirals are currently useful for prophylaxis or treatment of influenza?

A: Oseltamivir and zanamivir. Because of the emergence of drug-resistant strains, amantadine and rimantadine are not recommended in the United States.

Q: If a patient has a patchy infiltrate on a chest x-ray and bullous myringitis, what antibiotic should be prescribed?

A: An antibiotic such as erythromycin or a tetracycline for *Mycoplasma pneumoniae*

Q: What is the antibiotic of choice for community-acquired aspiration pneumonia?

A: Either penicillin or clindamycin, but metronidazole + penicillin or amoxicillin/clavulanic acid are good alternatives

Q: What is the drug of choice for streptococcal pharyngitis?

A: Penicillin is the drug of choice for infections due to *Streptococcus pyogenes*. For gram-positive cocci in general, penicillin G or V are always good guesses. Sometimes, such as for the enterococci, combination therapy may be required. In most cases, you will not need to know combination therapy except for a few bacteria such as *Mycobacterium tuberculosis* and *Helicobacter pylori*.

Q: What are the drugs of choice for *Mycobacterium tuberculosis*?

A: Isoniazid, rifampin, pyrazinamide, and ethambutol for 6–9 months is a recommended primary drug combination. Second-line drugs include cycloserine, ethionamide, p-aminosalicylic acid, streptomycin, capreomycin, amikacin/kanamycin,* levofloxacin,* moxifloxacin,* and gatifloxacin* (* = not FDA approved for tuberculosis). You should also be aware of any toxicities associated with these (and other) drugs, because these are covered on the pharmacology section of standardized exams.

Q: What is the cause of Hansen disease and what is the antimicrobial therapy?

A: Hansen disease (leprosy) is due to *Mycobacterium leprae*. The primary drug is dapsone and is coadministered with rifampin. (Trivia: Hansen disease is named for Norwegian scientist Gerhard Armauer Hansen, who first suggested and then demonstrated that leprosy was caused by a microbe.)

Q: What is the mode of action of dapsone?

A: It is the same as for the sulfa drugs, that is, it is a competitive inhibitor of para-aminobenzoic acid, blocking the synthesis of folic acid in bacteria.

Q: Polymyxin B has a high toxicity if used systemically. Why?

A: The polymyxins cause membrane leakage, not only in bacterial cells but also in human cells. They are bactericidal.

Q: What is the spectrum of antimicrobial activity of linezolid and quinupristin/dalfopristin?

A: They are antibiotics that have been FDA approved for use in infections due to vancomycin-resistant enterococci. Linezolid is an oral oxazolidinone with activity against *Enterococcus faecium* and *E. faecalis*. It is also active against MRSA and penicillin-resistant *Streptococcus pneumoniae*. The combination of quinupristin and dalfopristin (both are streptogramins) acts synergistically. It is indicated for bacteremia due to vancomycin-resistant *E. faecium*, but not *E. faecalis*. It can also be used in skin infections due to methicillin-susceptible *Staphylococcus aureus* or *Streptococcus pyogenes*.

Q: What is the therapy for *Helicobacter pylori*?

A: A combination of a proton pump inhibitor + clarithromycin + either metronidazole or amoxicillin. An alternative is bismuth subsalicylate + metronidazole + tetracycline + either a proton pump inhibitor or H₂ blocker.

Q: What is the mode of action of isoniazid? Rifampin? Pyrazinamide?

A: Isoniazid inhibits mycolic acid synthesis in the cell walls of mycobacteria. Rifampin inhibits RNA synthesis by binding to DNA-dependent RNA polymerase. The mode of action of pyrazinamide is unclear and therefore you do not need to know it.

Q: What is the mode of action of tetracycline?

A: It inhibits protein synthesis by binding to the 30S ribosomal subunit and preventing the aminoacyl-tRNA from attaching to the subunit. The tetracyclines are bacteriostatic.

Q: What is tigecycline?

A: It is a broad-spectrum antibiotic that has activity against a broad range of gram-positive and gram-negative bacteria, atypical and anaerobic bacteria, and antibiotic-resistant species, including MRSA, vancomycin-resistant *Enterococcus*, and penicillin-resistant *Streptococcus pneumoniae*. Tigecycline is the first FDA-approved antibiotic in a new class called glycylcyclines. Although it is a glycylcycline, its structure resembles the tetracyclines (it is a derivative of minocycline). Its mode of action is also similar to the tetracyclines (protein synthesis inhibitor by binding to the 30S ribosomal subunit and blocking the entry of amino-acyl transfer RNA into the acceptor site). An advantage of the glycylcyclines is that they overcome two types of genetic mechanisms responsible for tetracycline resistance: active efflux and ribosomal protection.

Q: What is the mode of action of metronidazole?

A: After transport into the cell, metronidazole is reduced to its active form (by an enzyme found in both anaerobes and some parasites) that results in the production of highly reactive free radicals. These cause irreversible damage to the DNA, resulting in cell death. If this was a multiple choice question, the answer would be that it inhibits DNA synthesis.

Q: A 47-year-old man presented to the hospital emergency department with a urethral exudate and complained of some discomfort on urination. Gram stain of the exudate revealed gram-negative diplococci within neutrophils. What is the treatment?

A: The patient probably has gonorrhea. The drug of choice for *Neisseria gonorrhoeae* is frequently ceftriaxone, but other drugs are also effective, such as cefixime and some fluoroquinolones. Because there is a high probability that he is coinfected with chlamydia, he should also be treated with azithromycin or doxycycline. This question gives me an opportunity to help you in taking exams. Be sure to read questions carefully. In this question, it mentions "gram-negative diplococci." Although you may read it correctly, your brain may register "gram-*positive* diplococci." Before you answer exam questions, make sure you know exactly all the details in the stem (i.e., don't read too fast).

Q: What is ceftriaxone and what is its mode of action?

A: It is a third-generation cephalosporin. All cephalosporins are beta-lactam antibiotics that have the same mode of action as the penicillins, namely, inhibition of peptidoglycan synthesis by binding to, and thus inactivating, the transpeptidase. As with all beta-lactams, cephalosporins are bactericidal.

Q: What is azithromycin and what is its mode of action?

A: It is a macrolide. Macrolides inhibit protein synthesis by binding to the bacterial 50S ribosomal subunit and blocking translocation, peptide bond formation, and release of the tRNA (with attached incomplete peptide). Erythromycin and clarithromycin are other macrolides.

Q: What are the resistance mechanisms of bacteria to the macrolides?

A: There are three common mechanisms of resistance: alteration in the drug target site (rRNA) by methylation, thus preventing the drug from binding; efflux of the antibiotic from the bacteria; and drug inactivation.

Q: Describe the mode of action of tetracycline and the mechanism of bacterial resistance.

A: Tetracycline and its derivatives inhibit protein synthesis by inhibiting the binding of aminoacyl-tRNA to the "A" (acceptor) site on the 30S ribosomal subunit. Resistant bacteria actively pump tetracycline back out of the cell (efflux). There are other mechanisms of resistance, such as enzymatic inactivation and the synthesis of ribosomal protection proteins, but these are not as important as efflux.

Q: Large "boxcar"-shaped gram-positive anaerobic bacilli were isolated from a postsurgical infection. The bacteria were subsequently shown to be obligate anaerobes that demonstrated a double zone of hemolysis on blood agar plates. What is the drug of choice?

A: Penicillin G. The organism is most likely *Clostridium perfringens*. Clindamycin or metronidazole is a reasonable alternative if penicillin is not a choice.

Q: What is imipenem and what is significant about its administration?

A: Imipenem is a beta-lactam (a carbapenem) antibiotic. It is given with cilastatin, which inhibits its inactivation in the kidneys. It has the broadest range of activity of all the beta-lactam antibiotics and, because there can be some significant toxicity, is usually reserved for severely ill patients. Meropenem is a less toxic carbapenem and does not require the coadministration of cilastatin.

Q: Cerebrospinal fluid from a 5-year-old child was Gram stained (after being concentrated by centrifugation). Gram-negative diplococci were observed. What is the drug of choice?

A: Penicillin G. The organism is probably *Neisseria meningitidis*.

Q: What is the treatment for tetanus?

A: Penicillin G, human tetanus immunoglobulin, muscle relaxants, tetanus toxoid, and supportive care. The purpose of the penicillin is to prevent any further growth of the bacteria and subsequent synthesis of the toxin.

Q: What is the therapy for antibiotic-associated pseudomembranous colitis?

A: Withdraw the drug that led to the disease or treat with oral metronidazole (first choice) or vancomycin. The organism is *Clostridium difficile*.

Q: What is the drug of choice for anthrax?

A: Ciprofloxacin or tetracycline. Penicillin G is a good alternative, especially for naturally acquired (non-weaponized) strains. The organism is *Bacillus anthracis*. As mentioned before, if you need to guess therapy for gram-positive bacteria, choose penicillin if it is a choice.

Q: What is the drug of choice for community-acquired urinary tract infection?

A: Trimethoprim-sulfamethoxazole. The most probable cause is *E. coli*.

Q: *E. coli* is frequently multiresistant. How do you choose empiric therapy for non–urinary tract infections when an antibiotic is indicated?

A: Choose a third-generation cephalosporin first, but in the meantime the organism should be tested in the lab for antimicrobial susceptibilities.

Q: What is the drug of choice for bacterial vaginosis?

A: Metronidazole (clindamycin if pregnant). Bacterial vaginosis is most likely a mixed infection that appears to involve *Gardnerella vaginalis*, *Mobiluncus*, and up to 17 additional (mostly anaerobic) bacteria.

Q: What is the drug of choice for meningitis due to *Haemophilus influenzae* type b?

A: A third-generation cephalosporin such as cefotaxime or ceftriaxone

Q: Why are trimethoprim and sulfamethoxazole frequently used in combination?

A: The combination is synergistic and resistance is less likely to develop.

Q: How does trimethoprim work?

A: It is an antimetabolite that preferentially binds to bacterial dihydrofolate reductase, thereby preventing the conversion of dihydrofolic acid into tetrahydrofolic acid.

Q: What is the mechanism of resistance to the sulfonamides and trimethoprim?

A: You probably do not need to know the resistance mechanisms for the sulfonamides because there are several (e.g., decreased permeability, new dihydropteroate synthetase, enzymatic inactivation). Trimethoprim resistance is primarily due to a new dihydrofolate reductase that will not bind the trimethoprim. Sulfonamides have no effect on human cells because we get preformed folic acid in our diet. Trimethoprim binds much better to microbial dihydrofolate reductase than to the human enzyme.

Q: What is the treatment for a patient (a volunteer for Doctors Without Borders) who developed a profuse watery diarrhea (resembling "rice water") who just returned from a trip to Luanda, Sudan?

A: The drug of choice is tetracycline because the diarrhea is most likely due to *Vibrio cholerae* (rice-water stools), but it is most important to treat with fluids and electrolytes.

Q: A 6-year-old boy recently returned from a family camping vacation in South Carolina. He became febrile and complained of a headache and muscle pain. A macular rash was noted on his extremities and included his palms and soles. What is the drug of choice?

A: Doxycycline. The infection is most likely Rocky Mountain spotted fever due to *Rickettsia rickettsii*. Alternative drugs include chloramphenicol or a fluoroquinolone. It might be helpful to remember that cell wall active antimicrobials are not effective against some of the atypical bacteria (rickettsia, chlamydia, and *Chlamydophila* are intracellular parasites and mycoplasmas have no cell walls).

Q: What is the drug of choice for amebiasis (amebic dysentery) and what is the etiological agent?

A: Mild to severe infection with *Entamoeba histolytica* can be treated with metronidazole (best guess) or tinidazole. There are other choices under certain circumstances, but you probably do not need to know them.

Q: What antimicrobials are used for malaria?

A: Chloroquine is the drug of choice for patients acquiring the disease in Central America (west of the Panama Canal), the Dominican Republic (not to be confused with Dominica), Haiti, and most of the Middle East. *Plasmodium falciparum* from these areas are generally chloroquine sensitive. For chloroquine-resistant strains of *P. falciparum*, there are several possibilities with various combinations, but the use of quinine is often part of the therapy. Specific treatment varies and depends on factors such as the infecting species, area where the infection was acquired, drug resistance, clinical status of the patient, any accompanying illnesses or conditions, pregnancy, drug allergies, and other medications taken by the patient.

Q: What is the drug of choice for *Pneumocystis* pneumonia?

A: Trimethoprim-sulfamethoxazole. Other drugs include dapsone, dapsone + pyrimethamine + leucovorin, atovaquone, or aerosolized pentamidine.

4 General Medical Microbiology

Q: For which bacteria (note: "bacteria" is plural, "bacterium" is singular) is the Gram stain *not* useful?

A: Mycoplasma (small size, no cell wall), chlamydia and rickettsia (intracellular), many spirochetes (thin), *Mycobacterium* (acid-fast), and *Legionella*. Most gram-positive bacteria that are old, dead, or damaged with antimicrobials can appear as gram-negative cells. Gram-positive bacteria under these circumstances are "leaky" and therefore appear red or pink rather than blue. The term "gram variable" refers to gram-positive bacteria that when Gram stained appear as a mixture of both gram-positive and a substantial number of gram-negative cells. The term is generally reserved for those species of gram-positive bacteria that routinely appear as both gram-positive and -negative cells when newly grown cells are stained. Almost all Gram stains of gram-positive bacteria contain a few cells that appear red, most likely because of a very small percentage of dead, thus leaky, cells.

Q: A 2-week-old neonate presents with clinical features suggestive of acute bacterial meningitis. Empiric antimicrobial therapy should include coverage for which bacteria?

A: *Escherichia coli*, *Streptococcus agalactiae*, and *Listeria monocytogenes*

Q: What pathogen does a "strawberry" cervix suggest?

A: *Trichomonas vaginalis*

Q: A lumbar puncture was performed on a 21-year-old college student who presented with symptoms of bacterial meningitis. Culture on blood agar resulted in the growth of small colonies that were alpha-hemolytic. Gram stain showed gram-positive lancet-shaped diplococci. What is the major virulence factor for this organism?

A: Capsule. This is a partial description of *Streptococcus pneumoniae*.

Q: A resident in family medicine asked a group of fourth-year medical students the following question: "The clinical bacteriology laboratory isolated a beta-hemolytic gram-positive streptococcus from the CSF of one of my patients. Is my patient most likely a neonate, 2 years old, or a college student?"

A: A neonate. The organism is most likely *Streptococcus agalactiae* (group B *Streptococcus*), one of the most common causes, along with *E. coli*, of meningitis in neonates. The key phrase is "most likely." The two most common beta-hemolytic gram-positive streptococci in human infections are *S. agalactiae* and *Streptococcus pyogenes* (group A strep). Both of these have the potential to cause a cerebrospinal fluid infection in a neonate, a 2-year-old, or a college student. *S. pyogenes* is not a common cause of cerebrospinal fluid infections in any age group, but it can occur (i.e., not likely).

Q: List the Centers for Disease Control and Prevention Category A bioterrorism diseases and their corresponding organisms.

A:

Anthrax: *Bacillus anthracis*

Botulism: *Clostridium botulinum* toxin (note: the toxin is the agent, not the organism)

Tularemia: *Francisella tularensis*

Plague: *Yersinia pestis*

Hemorrhagic fever: Arenaviruses (Lassa fever, lymphocytic choriomeningitis, Junin virus, Machupo virus, Guanarito virus; Bunyaviruses (Hantaviruses, Rift Valley fever); Flaviruses (Dengue); Filoviruses (Ebola, Marburg)

Smallpox: Variola major virus

Q: As a food, ham can be quite salty. What organism is most likely to be responsible for ham-associated food poisoning?

A: *Staphylococcus aureus*, which is a facultative halophile (grows in the presence of higher salt concentrations)

Q: What are three important genera of organisms that are acid-fast?

A: *Mycobacterium*, *Nocardia*, and *Cryptosporidium*. *Nocardia* species are usually referred to as being partially acid-fast, which means they do not decolorize well with weak solutions of acid. *Cryptosporidium* is a protozoan parasite.

Q: Where would a 20-year-old male patient most probably acquire a nosocomial infection?

A: In the hospital

Q: What major products of bacterial origin cause fever in a patient by inducing interleukin-1 production from macrophages?

A: Lipopolysaccharide and superantigens

Q: List at least three microbial causes of watery diarrhea.

A: This is a list of the some of the major organisms. It is not a complete list.
Escherichia coli: Enterotoxigenic strains cause travelers' and infant diarrhea.
Vibrio cholerae: rice-water stools; may be greenish due to mucus; more severe than others
Clostridium perfringens: rapid onset/short duration
Salmonella: may be bloody
Bacillus cereus: associated with reheated rice (another strain of *B. cereus* causes vomiting)
Clostridium difficile: antimicrobial-associated diarrhea and pseudomembranous colitis
Giardia: can be prolonged; severe, foul smelling, nonbloody
Cryptosporidium: watery diarrhea, immunocompromised (e.g., AIDS) at higher risk
Rotaviruses: especially in young children; viral diarrheas are not bloody
Norovirus (previously Norwalk or Norwalk-like virus): both diarrhea and vomiting; more common in older age groups (cruise ships, work places, etc. are frequent sites of outbreaks)
Adenoviruses: especially young children

Q: What is the advantage of antigenic (phase) variation to pathogens?

A: It allows them to evade immune surveillance.

Q: List some examples of organisms (bacteria, viruses, or parasites) that demonstrate antigenic variation.

A: *Neisseria gonorrhoeae* (pili and outer membrane proteins), *Salmonella* (pili), *Borrelia recurrentis* (outer membrane proteins), influenza virus (hemagglutinin and neuraminidase), and the trypanosomes (outer glycoprotein) are some of the most notable. Others include *Bordetella pertussis*, *E. coli*, *Haemophilus influenzae*, *Helicobacter pylori*, *Streptococcus pneumoniae*, and *Plasmodium* (malaria).

Q: What is tetanospasmin?

A: It is the neurotoxin of *Clostridium tetani*. It is most commonly referred to as the tetanus toxin.

Q: What is the mode of action of the tetanus toxin?

A: It prevents the release of the inhibitory neurotransmitters glycine and gamma-aminobutyric acid, resulting in convulsive muscle contractions, leading to paralysis.

Q: What is the mode of action of the diphtheria exotoxin?

A: It inhibits protein synthesis by ADP ribosylation of elongation factor 2 (EF-2). EF-2 functions in protein synthesis. Ribosylation of EF-2 inactivates it so that it can no longer function in protein synthesis. There are other bacterial exotoxins that function by ADP ribosylation of a target molecule. For example, *Pseudomonas aeruginosa* toxin A, although a different protein, has the same mode of action as the diphtheria toxin. The *E. coli* heat labile toxin and the toxins of cholera and pertussis cause ADP ribosylation of G proteins.

Q: What specific part of endotoxin has biological activity?

A: Lipid A, but we usually speak of the outer membrane (consisting, in part, of the lipopolysaccharide) of gram-negative bacteria as the endotoxin, and it is the endotoxin that has biological activity (even though we know it is really the lipid A portion of the outer membrane that has the biological activity).

Q: List some of the important biological activities of endotoxin.

A: Fever production, activation of classical and alternate complement pathways, shock, disseminated intravascular coagulation, and bone marrow suppression. There are others, but these are the major ones. Although not technically true, we often consider (1) the cell wall of gram-negative bacteria, (2) endotoxin, and (3) lipopolysaccharide as being synonymous, because lipid A is in each of these. As a side note, it should be mentioned that not all endotoxins have the same structure or degree of biological activity. For example, the endotoxin of *Bacteroides fragilis*, a gram-negative anaerobe, is chemically different from the endotoxins of the family Enterobacteriaceae (*E. coli*, *Salmonella*, etc.) and has little of the biological activity of classical endotoxins. There are many other examples of this in other species.

Q: With respect to the induction of cytokine production by endotoxin, what cells are induced to produce these endogenous substances?

A: Macrophages should be your first choice, but other cells, such as monocytes, endothelial cells, and epithelial cells, can be induced.

Q: What kinds of bacteria (gram-positive or gram-negative) produce exotoxins and what produces endotoxins?

A: Both gram-positive bacteria (e.g., *Clostridium tetani*) and gram-negative bacteria (e.g., *Vibrio cholerae*) can produce exotoxins. Only gram-negative bacteria produce endotoxins (by synthesis of lipopolysaccharide in the outer cell wall membrane).

Q: Fever production is usually associated with which type of toxin: exotoxin or endotoxin?

A: Endotoxin, but there are exceptions. The major exceptions are the superantigens of some bacteria such as the toxic shock syndrome toxin (known as TSST-1) and the exfoliative toxin (staphylococcal scalded skin syndrome) of *Staphylococcus aureus* and the toxins of *Streptococcus pyogenes* that are involved in streptococcal toxic shock syndrome and rheumatic fever. Superantigens are often secreted by bacteria, just like classical exotoxins such as the botulism and diphtheria toxins.

Q: What are the compositions of exotoxins and endotoxins?

A: Exotoxins are protein (as are superantigens). Endotoxins consist of the lipopolysaccharide of the outer membrane of gram-negative bacteria. It is the lipid A portion of lipopolysaccharide that is toxic.

Q: Which type of toxin usually can be easily inactivated by heat (heat labile): exotoxins or endotoxins?

A: Exotoxins

Q: With respect to the last question, there are some exotoxins that are heat stable. Name two.

A: The enterotoxin of *Staphylococcus aureus* and the heat stable enterotoxin of *E. coli*. Because of this heat stability, reheating food in which *S. aureus* has grown will not make it safe to eat. This does not generally apply to *E. coli*–contaminated food because the toxin is secreted in the intestinal tract rather than in the food.

Q: Can antitoxins be produced against both exotoxins and endotoxins?

A: Generally, antitoxins (protective antibodies) are produced against exotoxins, but not endotoxins. Antibodies are produced against endotoxins, but they are not protective. In other words, endotoxins are antigenic but not immunogenic.

Q: What kind of toxins can be converted into a toxoid?

A: Exotoxins, but not endotoxins

Q: What is a toxoid?

A: A toxoid is an inactivated toxin. Toxins can be inactivated by various methods, commonly by using heat or a chemical. Toxoids, therefore, have lost their toxicity but retain their immunogenicity. Toxoids are frequently used in bacterial vaccines.

Q: List some toxoids.

A: Diphtheria and tetanus toxoids are the most notable. Both are used in the DTaP, DT, Tdap, and Td vaccines.

Q: Describe the Gram stain morphology of *Neisseria*.

A: Gram-negative cocci in pairs with their flat sides facing each other (kidney bean shaped)

Q: In a suspected case of food poisoning, the clinical lab isolated a gram-negative rod that did not ferment lactose, was nonmotile, and did not produce hydrogen sulfide. What is the most likely pathogen?

A: *Shigella*. *Salmonella* and *Campylobacter* are both motile. Fermentation of lactose is usually only used to separate *Salmonella* and *Shigella* (both are nonlactose fermenters) from *E. coli* (lactose fermenter, i.e., lactose positive). *Salmonella* is hydrogen sulfide positive (turns iron-containing media, such as triple sugar iron, black).

Q: How does *Leptospira interrogans* typically infect humans?

A: Most commonly through direct or indirect contact with the urine of domestic animals and wild rodents. Reservoirs include deer, squirrels, foxes, and skunks, among others.

Q: What causes granuloma inguinale?

A: *Klebsiella granulomatis* (previously *Calymmatobacterium granulomatis*). If there is a question about this disease on a standardized exam (very unlikely), you most likely need to know only that it is a sexually transmitted disease and *maybe* be able to recognize the name of the organism.

Q: A chancre is observable in what stage or stages of syphilis?

A: Primary syphilis only

Q: A gram-positive coccus isolated from a furuncle (boil) was found to be catalase and coagulase positive. What is the most likely identification?

A: *Staphylococcus aureus*. For most examinations, this is the only gram-positive coccus that is both catalase and coagulase positive. All the streptococci and enterococci are catalase negative.

Q: In which age group would you expect to find the highest incidence of *Haemophilus influenzae* type b infection?

A: Approximately 3 months to 3 years. Because of the use of the Hib vaccine in developed countries, the incidence of *H. influenzae* type b infection has decreased more than 10-fold. Meningitis and sepsis due to this organism in this age group have obviously decreased, whereas similar infections due to other bacteria such as *Streptococcus pneumoniae* now cause a significant percentage of these infections. The infection rate due to other bacteria has not necessarily increased. In emerging countries where the Hib vaccine is not in wide use, *H. influenzae* type b is still a leading cause of bacterial pneumonia deaths in children. *H. influenzae* type a is a rare pathogen that appears to be emerging in some areas of the world. Some cases have been reported in the United States. It has a similar age distribution as type b.

Q: What is the most common cause of otitis media?

A: *Streptococcus pneumoniae* followed by *Haemophilus influenzae* and *Moraxella catarrhalis*. These are also common causes of bacterial upper respiratory tract infections.

Q: How does *Listeria monocytogenes* evade the immune response?

A: It is a facultative intracellular parasite that can invade macrophages. It can then spread from macrophage to macrophage without ever getting into the bloodstream. As a result, it is sequestered with respect to the humeral immune response.

Q: Almost all encapsulated bacterial pathogens have a capsule that is polysaccharide. What pathogen does not have a *polysaccharide* capsule?

A: *Bacillus anthracis* has a capsule composed of poly-glutamic acid.

Q: What is the most common cause of aspiration pneumonia?

A: Mixed anaerobes aspirated from the oral cavity, but aspiration of stomach contents is also a source

Q: What pathogen shows a characteristic tumbling motility at room temperature when viewed by dark-field microscopy (wet mount)?

A: *Listeria monocytogenes*

Q: What organism is usually involved and what heart valve is *most commonly* affected in an intravenous drug abuser?

A: The tricuspid valve, usually with *Staphylococcus aureus*

Q: What is the most probable cause of postburn sepsis due to a gram-negative oxidase-positive bacillus that produces a bluish-green pigment on culture?

A: *Pseudomonas aeruginosa* (Don't you wish they were all this easy!)

Q: What is the most common cause of community-acquired pneumonia in adults? In children?

A: *Streptococcus pneumoniae* in adults (especially the elderly or persons with chronic obstructive pulmonary disease) and children. Respiratory syncytial virus is the most common viral pathogen in children.

Q: What antimicrobials should be used for the initial management of community-acquired pneumonia?

A: This question was easier to answer (penicillin) before the emergence of drug-resistant strains of *S. pneumoniae* and other pathogens. Therapy must also take into account patient characteristics (inpatient, outpatient, age, comorbidities such as chronic obstructive pulmonary disease, recent treatment that could have selected for antibiotic resistant strains, etc.). Resistance of *S. pneumoniae* to the beta-lactams and the macrolides is common.

Up to 30% of isolates in some areas of the country are resistant to these. If there are no significant risks for drug-resistant strains, a macrolide or doxycycline are reasonable choices (these also are effective against *Mycoplasma pneumoniae* and *Chlamydophila pneumoniae*). Fluoroquinolones and cephalosporins may also be good as empirical treatment. Combination therapy such as a beta-lactam + a fluoroquinolone or macrolide should be considered in severe cases. Therapy can be modified after susceptibility results become available.

Q: Sputum that is the color and consistency of currant jelly is indicative of pneumonia due to what organism?
   A: *Klebsiella pneumoniae*
   Trivia: Currants are small berries (related to gooseberries) from ornamental shrubs that grow in most climatic zones. Although red is a popular color, there are also black, white, and pink varieties. Most commonly used to make jelly or pies, they can also be used in wine making. Folk medicine uses, especially the fruit and leaves from black currants, include colds, flu, tuberculosis, miscarriage, dermatitis, and sore eyes, among others.

Q: What are some common predisposing conditions of staphylococcal pneumonia?
   A: Drug use, influenza, hospitalization (these patients are commonly immunosuppressed), and endocarditis are some. The important point here is that *S. aureus* rarely causes a primary pneumonia without predisposing factors. This pneumonia produces high fever, chills, and a purulent productive cough.

Q: A 16-year-old boy comes into your office with a dry cough, malaise, mild fever, and a sore throat that have developed in the past month. What is the diagnosis?
   A: Mycoplasma pneumonia. This condition usually has a slow onset and occurs in the young. Treatment is with a tetracycline or a macrolide such as erythromycin.

Q: A 2-year-old nonimmunized child was suspected of having bacterial meningitis. Cerebrospinal fluid collected by lumbar puncture was sent to the clinical bacteriology lab for culture. What medium should be used to optimize the isolation of the gram-negative organism that, when it causes meningitis, would most likely infect a child of this approximate age?
   A: Chocolate agar supplemented with X (hemin) and V (NAD) factors for the isolation of *Haemophilus influenzae*. When *H. influenzae* causes meningitis, it is usually in the age group of 3 months to 3 years. You will also see this organism in immunosuppressed individuals including the elderly with waning immune systems. *Streptococcus pneumoniae* is the most common cause of bacterial meningitis in all except neonates.

Q: What is the most common nosocomial illness?
   A: Urinary tract infection

Q: What is the second most common nosocomial illness?
   A: Pneumonia. Patients on mechanical ventilation are at especially high risk. Nosocomial bacteria is also common.

Q: What is the most common cause of nosocomial pneumonia?
   A: Gram-negative bacilli, especially *Pseudomonas aeruginosa*. Other common gram-negative etiologies include *Klebsiella*, *Enterobacter*, *Serratia*, and *Acinetobacter*. There is a low probability of *Serratia* and *Acinetobacter* appearing as a correct answer on a standardized exam, although *Acinetobacter* infections are increasing. Nosocomial pneumonias due to *Staphylococcus aureus* (gram-positive) are also common.

Q: Describe the classic chest x-ray findings in a patient with mycoplasma pneumonia.
   A: Patchy diffuse densities involving the entire lung are most common.

Q: A 16-year-old boy presents to your office complaining of pleurisy, sudden onset of fever and chills, and rust-colored sputum. What is the diagnosis?
   A: Pneumonia caused by *Streptococcus pneumoniae*. It is the most common community-acquired pneumonia, presenting as a consolidating lobar pneumonia. Hemoptysis is common. Pneumonia due to *S. pneumoniae* in an otherwise healthy 16-year-old would be unexpected, but it does occur.

Q: A 20-year-old college student is home for winter break and comes to your office complaining of a 3-week history of a nonproductive dry hacking cough, malaise, a mild fever, and no chills. What is the diagnosis?
   A: Atypical pneumonia (a walking pneumonia). *Mycoplasma pneumoniae* is one of the most common causes. Others include *Chlamydophila* (formerly *Chlamydia*) *pneumoniae*, *Chlamydophila* (formerly *Chlamydia*) *psittaci* (psittacosis; inhalation of dried bird feces), *Legionella pneumophila* (Pontiac fever, but also causes more serious legionellosis, most commonly in older adults, especially those with comorbidities such as chronic obstructive pulmonary disease), and *Coxiella burnetii* (Q fever, rare in the United States).

Q: Describe the different presentations of bacterial and viral pneumonia.

A: Bacterial pneumonia is typified by a sudden onset of symptoms, including pleurisy, fever and chills, productive cough, tachypnea, and tachycardia. Viral pneumonia is characterized by gradual onset of symptoms, no pleurisy, chills or high fever, general malaise, and a nonproductive cough.

Q: What is the most common cause of pneumonia in children?

A: Viruses, including influenza, parainfluenza, respiratory syncytial virus, and adenoviruses

Q: What is the most common cause of pulmonary infection in persons with cystic fibrosis?

A: *Pseudomonas aeruginosa* (most common), but *Staphylococcus aureus* (especially) and *Haemophilus influenzae* can initially be present, particularly in younger patients

Q: What community-acquired pneumonia is due to multiple organisms?

A: Aspiration pneumonia, due to bacteria, especially anaerobes, aspirated from the oral cavity

Q: A person with sickle cell anemia is at an increased risk of infection with what organisms?

A: Repeated episodes of splenic sequestration (pooling) eventually leads to significant damage to the spleen, and it becomes nonfunctional. This results in susceptibility to primarily encapsulated bacteria such as *Streptococcus pneumoniae*, *Haemophilus influenzae* type b, and *Neisseria meningitidis*. Other impairments to the immune system cause increased susceptibility to other pathogens, including *Salmonella* (osteomyelitis), *Staphylococcus aureus*, *E. coli*, and *Klebsiella*. Human parvovirus B19 can cause a temporary pause in erythrocyte production (aplastic crisis).

Q: A person with sickle cell anemia is at a decreased risk of infection with what organism?

A: *Plasmodium* species, the etiological agents of malaria

Q: List some oncogenic viruses and their associated cancers.

A: EBV, Burkitt lymphoma and nasopharyngeal carcinoma; HBV, hepatocellular carcinoma; HHV 8, Kaposi sarcoma; HPV, cervical carcinoma; HTLV-1, adult T-cell leukemia

Q: What is the cause of Whipple disease?

A: It is caused by a bacterium called *Trophermyma whippleii* that is in the family Actinomycetales. (It is unlikely that this disease will be on any qualifying exam. If anything, you only need to know that it has a bacterial etiology.) Arthritis, diarrhea, weight loss, and lymphadenopathy characterize Whipple disease. It is an extremely rare disease of primarily middle-aged white males.

Q: The most common cause of gastritis is what?

A: *Helicobacter pylori*. Gastritis can lead to peptic ulcer disease and stomach cancer.

Q: Guillain-Barré syndrome (GBS) is an autoimmune disorder directed to the nervous system (demyelination). It may have a microbial etiology. For what infectious organisms has an association to GBS been suggested?

A: The most common predisposing condition is gastroenteritis due to *Campylobacter jejuni*. Other infectious etiologies include cytomegalovirus, Epstein-Barr virus, and *Mycoplasma pneumoniae*. Additional factors that may lead to GBS are vaccination (e.g., influenza, rabies, and meningococcal in past years, dependent on manufacturer), surgery, lymphoma, and systemic lupus erythematosus. All these pathogens have carbohydrate surface antigens similar to those in peripheral nerve tissue. *Campylobacter* is the most important association to remember. GBS is the most frequent cause of acute flaccid paralysis in humans (1–2 cases/100,000 people a year). It is an acute inflammatory polyradiculoneuropathy that usually develops after a gastrointestinal infection. Clinical symptoms frequently occur 1–3 weeks after a bacterial or viral infection.

Q: What is Reiter syndrome (aka reactive arthritis) and what microbial infection can lead to it?

A: The three most common symptoms of Reiter syndrome consist of the triad of urethritis, conjunctivitis, and arthritis. It is an autoimmune disease that has a strong association with HLA type B27. *Chlamydia trachomatis* infection is the most common initiating event. Gastrointestinal infection caused by *Salmonella*, *Shigella*, *Campylobacter*, or *Yersinia* may lead to Reiter syndrome. There can be other infectious etiologies.

Q: What is the most common cause of otitis media?

A: Bacteria cause most cases, most commonly *Streptococcus pneumoniae*, *Haemophilus influenzae*, and *Moraxella catarrhalis*, probably in this order. About 15% of cases of acute otitis media are due to viruses such as reparatory syncytial virus and rhinovirus.

Q: What is the recommended antimicrobial approach to otitis media in children?

A: Recent evidence suggests a "wait-and-see prescription" approach for antibiotic intervention. Parents are asked not to fill the prescription if the child's condition is not worsening or not improving within two days. Because otitis media historically is the most common infection in children that is treated with antibiotics, this approach will help in controlling the unnecessary use of antibiotics and lessen the selection for antibiotic resistance.

Q: What is lymphogranuloma venereum?

A: Lymphogranuloma venereum is a sexually transmitted invasive lymphatic infection caused by *Chlamydia trachomatis*. It is not as common in the United States as it is in some developing countries. Most cases in the United States are diagnosed in patients who have recently traveled to endemic areas.

Q: What is the treatment for tetanus?

A: Penicillin to prevent additional growth of bacteria, human antitetanus immunoglobulin to inactivate any circulating toxin, and supportive care (nutritional, respiratory, etc.). If the wound is obvious, surgical debridement to remove infected tissue (and to encourage re-oxygenation of the wound) may be indicated. Patients are usually given the tetanus toxoid and a muscle relaxant. Because of the possibility of spontaneous muscle contractions, patients should be kept in a dark, quiet environment.

Q: What microscopic findings should be present to consider a sputum sample adequate for further workup?

A: Finding more than 25 polymorphonuclear neutrophilic leukocytes (usually abbreviated as PMNs) and less than 10 squamous epithelial cells per low power field (100× total magnification)

Q: What organisms are involved in pleuropulmonary infections after aspirating oropharyngeal material?

A: Primarily anaerobes, resulting in infections such as necrotizing pneumonia, lung abscess, "typical" pneumonia, and empyema. Because nonanaerobes, such as *Streptococcus pneumoniae*, *Staphylococcus aureus*, *Haemophilus influenzae*, and gram-negative enterics (family Enterobacteriaceae) are also commonly present, empiric therapy should not be directed only at anaerobes.

Q: A 6-month-old infant is constipated, flaccid, and only stares straight ahead. He is not running a fever. However, his mother is especially worried because he will not take his bottle, not even with the honey that the mother usually puts on the tip of the nipple. What is the diagnosis?

A: Infant botulism. Infants ingest spores of *Clostridium botulinum* that are in the environment, most commonly in honey. The spores germinate in the colon where the vegetative cells produce the toxin. It is unclear why infants are more susceptible to this type of botulism than older children or adults.

Q: How does infant botulism differ from food-borne botulism with respect to toxin production (i.e., the location of the bacteria that are secreting the toxin)?

A: Infant botulism is caused by toxin production *in the gastrointestinal tract* after ingestion and germination of the spores of *C. botulinum*. Food-borne botulism occurs after the ingestion of preformed toxin produced *in the food* after the spores germinate.

Q: What is the therapy for food-borne botulism?

A: Trivalent A-B-E antitoxin and supportive (especially respiratory)

Q: A 12-year-old boy complains of two days of rice-water stools, muscle cramps, and extreme fatigue. He looks pale, dehydrated, and very ill. The patient states that he has just returned from India with his father. What is the diagnosis?

A: Cholera. The rice-water stools are always a giveaway. In the United States cholera usually develops in travelers returning from endemic areas, such as India, Africa, Southeast Asia, southern Europe, and the Middle East. This disease can be prevented if the water ingested during the trip is purified (e.g., bottled). Untreated water, unpeeled fruits or vegetables, and seafood should be avoided altogether.

Q: Tabes (pronounced *tay beez*) dorsalis is sometimes seen in what infectious disease?

A: Tertiary syphilis. It is due to the demyelination of the dorsal columns in the spinal cord, which are normally responsible for the sense of position.

Q: How does tabes dorsalis present?

A: Loss of position sensation causes severe gait and leg ataxia (balance and motor control problems) and can result in a staggering wide-based gait, postural instability, pain, paresthesias, dementia, deafness, visual impairment, and impaired response to light. Tabes dorsalis can also be due to spinal cord injury in the absence of infection.

Q: Infections with what organisms are associated with Reye syndrome?

A: Young children infected with varicella-zoster virus or influenza virus (especially B, but sometimes A) are at risk when aspirin is used to control fever.

Q: In which diarrheal illnesses are fecal leukocytes visible by microscopy?

A: The most common are in diarrheas due to invasive organisms, such as *Shigella*, *Campylobacter*, and certain strains (e.g., enteropathogenic and enteroinvasive strains) of *E. coli*. Others include *Salmonella*, *Yersinia enterocolitica*, *Vibrio parahaemolyticus*, *V. vulnificus*, and *C. difficile*. Fecal white blood cells are absent in *Vibrio cholerae* (but can be present in non-cholera *Vibrio* spp.) and in viral and parasitic infections.

Q: Name the most probable cause of diarrhea in a 6-month-old in day care.

A: Viral diarrhea caused by rotavirus is the most common. Also, test for *Giardia* and *Cryptosporidium* because these agents are on the list of day care–associated diarrheal illnesses.

Q: What class of organisms (bacteria, viruses, parasites) is the most common cause of acute diarrhea?

A: Viruses. Viral diarrhea is generally self-limited, lasting only 1–3 days.

Q: Diarrhea that develops within 12 hours of a meal is most probably caused by what?

A: An ingested preformed toxin, such as the enterotoxin of *Staphylococcus aureus*. Other fast-onset/short-duration food poisonings are due to *Bacillus cereus* and *Clostridium perfringens*.

Q: Which hepatitis viruses can be transferred by blood?

A: Hepatitis B (HBV), C (HCV), and D (HDV) viruses. How to remember? If you write out the hepatitis viruses in order, A through E (do not worry about any newly described hepatitis viruses such as F or G), the three on the inside (B, C, and D) can be spread via blood (and mucous membrane contact). Blood is also on the inside (of us). How are hepatitis viruses A (HAV) and E (HEV) transmitted? By feces (which, when contagious, is on the outside) via the fecal–oral route.

Q: Which hepatitis viruses are enveloped?

A: HBV, HCV, and HDV. You can remember this by a similar method as in the previous question. The three viruses on the inside (B, C, and D) have their nucleic acid, capsid, and other structural parts on the inside of a membrane.

Q: Which hepatitis viruses can result in chronic infections?

A: By using the same method as described above, you can see that chronic infections stay on the inside, just like HBV, HCV, and HDV. These three viruses also have the highest death rates, due to chronic liver disease and associated cirrhosis or liver cancer.

Q: Which hepatitis viruses have DNA and which have RNA?

A: This question is not that difficult. All hepatitis viruses have RNA except HBV. It is a hepa**dna**virus that has DNA. HAV is a pico**rna**virus; HCV is a flavivirus. HEV was formally classified as a calicivirus, but it has been reclassified in a new family, the Hepeviridae, where it is currently the sole member.

Q: Which hepatitis virus will not replicate unless the host is coinfected with HBV?

A: HDV is defective and requires a helper virus (HBV). It will not replicate in a patient without an active HBV infection. It requires HBsAg for completion of the replication cycle. About 4% of patients with HBV are coinfected with HDV. HDV infection is most common in intravenous drug abusers who are also infected with HBV.

Q: How is HAV infection prevented in someone not yet exposed to the virus?

A: HAV can be prevented with a vaccine, with or without the administration of HAV immune globulin, depending on the circumstances. The immune globulin should be used for postexposure prophylaxis. Recovery from HAV infection results in lifelong immunity.

Q: How is HBV infection prevented in someone not yet exposed to the virus?

A: HBV can be prevented with a vaccine. Postexposure prophylaxis is with the administration of HBV vaccine and immune globulin.

Q: How is HCV infection prevented in someone not yet exposed to the virus?

A: This can be more difficult because there is no vaccine or immunoglobulin available. Prevention relies on a safe blood supply and behavior modification in those at risk for spreading the virus (intravenous drug abusers, recently tattooed/body pierced, unprotected sex with multiple partners or whose health status is unknown).

Q: Can HCV be perinatally transferred?

A: Yes, but only when the mother is positive for HCV RNA at the time of delivery. The risk of transfer is about 6% but can increase to nearly 20% if the mother is coinfected with HIV.

Q: Can HCV be transferred by breast-feeding?

A: Although theoretically possible, HCV transmission through breast-feeding has not been demonstrated in HCV–positive mothers. As a precautionary measure, mothers with cracked or bleeding nipples should refrain from breast-feeding.

Q: How is HDV infection prevented in someone not yet exposed to the virus?

A: This is a problem-solving question in disguise as rote memorization. Remember that HDV is defective and requires an active HBV infection for replication. Thus the HBV vaccine prevents HDV infection as well.

Q: How is HEV infection prevented in someone not yet exposed to the virus?

A: As with HCV, this is difficult because there is no vaccine or immunoglobulin available. Prevention relies on a safe water supply. Therefore HEV infection is rare in the United States and other countries that have appropriately treated water. In the United States it is associated with recent foreign travel to endemic areas. HEV infection is most common in Asia, China, and Mexico.

Q: Match the following hepatitis serologies with the correct clinical description.

(1) HBsAg (−), and anti-HBs (+)
(2) IgM HBcAg (+), anti-HBs (−)
(3) IgG HBcAg (+), anti-HBs (−)
(4) HBeAg (+)

(a) Ongoing viral replication, highly infectious
(b) Remote infection, not infectious
(c) Prior infection or vaccination, not infectious
(d) Recent or ongoing infection; a high titer means high infectivity, whereas a low titer suggests chronic, active infection

A: (1) c, (2) d, (3) b, and (4) a
HBeAg (+): "e" = "Eeek! I'm infectious!" for e ANTIGEN, anti-HBe implies decreased infectivity
IgG HBcAg (+): "G" = gone
IgM HBcAG (+): "M" = might be contagious still
anti-HBs (+): "s" = stopped. Patient has antibodies to surface antigen.

Note: You must specify whether the antigen or the antibody is present when learning these letter codes; otherwise, you will learn them the wrong way around. Go over it 10 times in your text of choice so it will make sense.

Q: Which hepatitis viruses are associated with hepatocellular carcinoma?

A: Hepatitis B virus (DNA), especially in areas of China and Africa, and hepatitis C virus (RNA; flavivirus)

Q: A 7-year-old boy had bloody diarrhea that lasted for eight days, abdominal cramps, vomiting, and a fever as high as 103.7°F. It was suspected that the boy became infected after handling the newly acquired pet iguana. What is the most likely pathogen?

A: *Salmonella*. Reptiles (turtles, iguanas, snakes, etc.) commonly are carriers of various serotypes of *Salmonella* that result in the same symptoms associated with food-borne disease.

Q: List the routine vaccines that are available in the United States for infants or children (≥18 years) who have no risk factors.

A: Hepatitis A and B (separate vaccines); diphtheria, tetanus, pertussis (DTaP); *Haemophilus influenzae* type b (Hib); poliovirus; rotavirus; measles-mumps-rubella (MMR); varicella, meningococcus, pneumococcus, human papillomavirus virus (HPV; females)

Q: For each vaccine, what is the composition?

A:

Hepatitis A: HepA is an inactivated virus.
Hepatitis B: HepB is composed of HBsAg (from genetically engineered yeast).
Diphtheria, tetanus, pertussis: DTaP; diphtheria and tetanus toxoids, acellular *Bordetella pertussis* components (exact composition varies with manufacturer; one type consists of pertussis toxoid, another pertussis toxoid and filamentous hemagglutinin; there are several other formulations including pertussis whole-cell vaccines [DPT])

Haemophilus influenzae type b: Hib; polysaccharide capsule conjugated to proteins such as a mutant dip~~~~
    toxin (nontoxigenic) or diphtheria toxoid, or outer membrane protein from *Neisseria meningitidis* ~~~~
    toxoid

Poliovirus: live attenuated virus (Sabin) or inactivated virus (Salk)

Rotavirus: live attenuated viruses (oral pentavalent vaccine)

MMR: live attenuated viruses

Varicella (chickenpox): live attenuated virus

Meningococcus: MCV4 (preferred) composed of *Neisseria meningitidis* serogroups A, C, Y, and W-135 capsular
    polysaccharide antigens individually conjugated to diphtheria toxoid protein; MPSV4 composed of purified
    capsular polysaccharide antigens from *N. meningitidis*, serogroups A, C, Y, and W-135 (nonconjugated)

Pneumococcus: PCV7 is a heptavalent capsular polysaccharide vaccine.

Human papillomavirus virus (HPV): noninfectious recombinant, quadrivalent vaccine prepared from the purified
    virus-like particles (VLPs) of the major capsid (L1) protein of HPV types 6, 11, 16, and 18; recommended for
    girls ages 11–12 (range, 9–26).

Q: A 4-year-old child complains to his mother that his "bottom" itches. What is the most common infectious cause of itching of the anus in children?

    A: *Enterobius vermicularis*, or pinworm

Q: How is pinworm infection diagnosed?

    A: The Scotch tape test is used for detection of the pinworm eggs (ova). Apply the sticky-side down to the perianal area and then smooth the tape out on a glass slide using a cotton swab. Examine the slide for eggs with a microscope set on low power.

Q: A patient developed diarrhea and abdominal pain after taking antibiotics for two weeks for an infection. What might a sigmoidoscopy reveal?

    A: Yellowish superficial plaques. This finding is indicative of pseudomembranous colitis.

Q: What is the most common infectious cause of antibiotic-associated pseudomembranous colitis?

    A: *C. difficile*. This can be confirmed by demonstration of either toxin A or B in the stools. Some cases of community-acquired pseudomembranous colitis may not be associated with antibiotic use. In hospitals, the use of protein pump inhibitors may be related to some cases in which antibiotics are not implicated. Complications of the colitis include perforation and peritonitis, leading to death.

Q: What is the treatment for pseudomembranous colitis?

    A: Stop antimicrobials (if possible) if the patient is taking any (~25% of diarrhea in patients will resolve) or give oral metronidazole (first choice) or vancomycin (if no response to metronidazole; several other factors, beyond the purposes of this publication, may necessitate vancomycin usage).

Q: Name the infectious causes of bloody stools (hematochezia).

    A: *Salmonella, Shigella, Yersinia enterocolitica* (children under 1 year old), *Campylobacter jejuni, Escherichia coli* (some strains, such as enterohemorrhagic *E. coli*), and the protozoan *Entamoeba histolytica* are the major ones you should know. Melena (black "tarry" stools) can be seen with peptic ulcers due to any cause, including *Campylobacter*, or in other noninfectious causes of gastrointestinal bleed. Viruses do not cause bloody stools.

Q: Two children from the same family present with severe nausea and vomiting about 4–6 hours after eating. What are the most likely causes?

    A: *Staphylococcus aureus* (enterotoxin), *Bacillus cereus* (emetic enterotoxin; there is also a diarrheal form of enterotoxin), or heavy metals. Ingestion of preformed toxins or poisons has a rapid onset.

Q: Distinguish between anal chancres of primary syphilis and herpetic ulcers.

    A: Anal chancres of primary syphilis are painful, symmetric, indurated, and diagnosed by dark-field microscopy (observation of the motile spirochetes). Conversely, herpes simplex produces perianal paresthesias and pruritus followed by red-haloed vesicles and ruptured vesicles. A Tzanck test, demonstrating multinucleated giant cells, is useful in the diagnosis of herpes simplex. There is also a simple direct fluorescent antibody test for herpes infections. Both infections cause painful inguinal adenopathy.

Q: You stick yourself with a needle from a chronic hepatitis B carrier. You have been vaccinated but have never had your antibody status checked. What is the appropriate postexposure prophylaxis?

    A: Measure your anti-HBs titer. If it is adequate, treatment is not required. If it is inadequate, you need a single dose of hepatitis B immunoglobulin and a vaccine booster.

Q: What drugs can be used for the treatment of chronic hepatitis B?

A: Adefovir dipivoxil, interferon alfa-2b, pegylated interferon alfa-2a, lamivudine, entecavir, and telbivudine

Q: What is the most common opportunistic infection in AIDS patients?

A: *Pneumocystis jiroveci* pneumonia, also called *Pneumocystis* pneumonia. (The previous name for the organism was *Pneumocystis carinii* pneumonia, abbreviated PCP. The abbreviation PCP is still used to designate *Pneumocystis* pneumonia.) Symptoms may include a nonproductive cough and dyspnea. A chest x-ray may reveal diffuse interstitial infiltrates, or it may be negative. Although gallium scanning is more sensitive, false positives occur. Initial treatment includes trimethoprim-sulfamethoxazole. Pentamidine is an alternative. Other common opportunistic pathogens include the *Mycobacterium avium* complex, *Toxoplasma* (encephalitis), cytomegalovirus, *Cryptococcus neoformans*, and *Candida* spp.

Q: What is the most common oral manifestation of AIDS?

A: Oropharyngeal thrush (oral candidiasis) due to *Candida albicans*. Some other AIDS-related oropharyngeal diseases are Kaposi sarcoma (HHV-8), hairy leukoplakia (EBV), and non-Hodgkin lymphoma (some).

Q: What is the most common cause of urethritis in males?

A: Urethritis is a sexually transmitted disease primarily caused by *Neisseria gonorrhoeae* (gonococcal urethritis) or *Chlamydia trachomatis* (nongonococcal urethritis). Gonorrhea presents with a purulent discharge from the urethra, whereas chlamydia is generally associated with a thinner, white mucous discharge. Other less common agents include *Ureaplasma urealyticum*, *Mycoplasma hominis*, *Mycoplasma genitalium*, *Trichomonas vaginalis* (especially older men), and herpes simplex virus type 2. In men who engage in insertive anal intercourse, enteric pathogens can be involved.

Q: How should urethritis in males be treated?

A: Because of the high incidence of coinfection, treatment should cover both gonorrhea and chlamydia. Ceftriaxone, cefixime, ciprofloxacin, ofloxacin, or levofloxacin is recommended for gonorrhea. Quinolone-resistant *N. gonorrhoeae* occurs in Asia, the Pacific Islands, and California. Therefore quinolones are no longer recommended for the empirical treatment of gonorrhea in these areas. For chlamydia, doxycycline, azithromycin, and ofloxacin are the drugs of choice.

Q: What is the cause of condylomata acuminata (genital warts)?

A: HPV types 6 and 11 are the most common subtypes of virus grown from these lesions. The risk of male-to-female transmission of HPV infection can be reduced with male condom use. Hopefully, the use of the HPV vaccine in females (recommended for ages 11–12 [range, 9–26]) will lead to fewer cases in both males and females in the future.

Q: What are the clinical manifestations of primary syphilis?

A: A painless genital ulcer (chancre) at the site of inoculation of the spirochete

Q: What are the clinical manifestations of secondary syphilis?

A: Maculopapular rash of the trunk, soles, and palms that can resemble pityriasis rosea; mucous patches (white patches on the mucous membranes); and condyloma latum, which are flat-topped warts present in moist areas of the body, such as the perineum

Q: What is the treatment for sexually acquired syphilis?

A: For primary syphilis, a single dose of intramuscular benzathine penicillin G is sufficient.

Q: What are the signs and symptoms of pelvic inflammatory disease, or PID?

A: Classic PID is characterized by fever and lower abdominal pain. The patient may have just finished her menses. Physical examination reveals fever, lower abdominal tenderness, and marked bilateral adnexal tenderness.

Q: What is the organism most often isolated from patients with PID?

A: *Neisseria gonorrhoeae* and *Chlamydia trachomatis* are the most common.

Q: What other microorganisms are responsible for PID?

A: Besides *N. gonorrhoeae*, other causes include the normal vaginal flora such as *Bacteroides* and *Peptostreptococcus* as well as mycoplasma and other bacteria.

Q: How is the definitive diagnosis of infection with *N. gonorrhoeae* made?

A: By incubating collected secretions on selective media, such as Thayer-Martin, in a 5–10% CO_2 environment and identifying the organism. This is required for the diagnosis in females. Presumptive diagnosis in males can be done microscopically by the observation of gram-negative diplococci within PMNs. Other rapid tests include DNA probes, polymerase chain reaction, and ligand chain reaction.

Q: There are two types of polio vaccine licensed for administration in the United States: oral and intramuscular. Which one carries a risk of vaccine-induced poliomyelitis?

A: The Sabin oral polio vaccine, which is a live attenuated vaccine. The intramuscular vaccine (Salk) is inactivated.

Q: What is the most important factor for reducing infectious disease transmission in hospitals?

A: Hand-washing

Q: What is the Jarisch-Herxheimer reaction?

A: It is a reaction consisting of headache, fever, myalgia, hypotension, and increased severity of disease symptoms, which occur after antibiotic treatment. It most commonly occurs in syphilis but can occur in other, especially spirochetal, infections (e.g., Lyme disease). For example, penicillin G causes the release of endotoxin from the treponemes as they die and lyse, resulting in symptoms similar to mild endotoxic shock. This is the Jarisch-Herxheimer reaction. The reaction may also result in neurological, auditory, or visual changes.

Q: What are some physical signs associated with infective endocarditis?

A: Splinter hemorrhages, Osler nodes, Janeway lesions, and Roth spots can be indications of infective endocarditis. Petechiae are frequent, but they are not too specific. You should review images of these before taking board exams.

Q: List some of the more common bacteria that cause infective endocarditis.

A: The most common of all is *Staphylococcus aureus*. Bacteria in the viridans group of streptococci (e.g., *S. mutans, S. mitis, S. salivarius,* and others) are responsible for about half of the cases of subacute endocarditis (there is *no* organism called *Streptococcus viridans*, although you will frequently see this listed). Other species of streptococci (groups A, B, C, D, and G), enterococci, and many gram-negatives can also be involved. Fungi, such as *Candida* and *Aspergillus*, are rare causes.

Q: Patients with diabetes mellitus should receive what two vaccines?

A: They should get an influenza vaccine annually and the pneumococcal vaccine (23 valent) before age 65 and a booster after age 65.

Q: Are elevated IgM levels in the fetus significant?

A: Yes. IgM is the only antibody that can be synthesized by the fetus. Therefore an increased level is indicative of either congenital or perinatal infection.

Q: Which of the hepatitis viruses are transmitted by the fecal–oral route?

A: Hepatitis A and E. Remember the earlier hint? List the five viruses in order: A, B, C, D, and E. A and E are on the *outside* of the other three. Hepatitis A and E come from *outside* (contaminated food/water), thus fecal–oral. The three remaining viruses are B, C, and D and are on the *inside* of A and E. Blood is on the *inside* of the body; therefore these three are blood-borne.

Q: A sexually active 16-year-old boy complains of dysuria and a penile discharge. What is the appropriate antimicrobial therapy of this patient?

A: You suspect urethritis. The two most common causes of urethritis in a sexually active male are *Chlamydia trachomatis* and *Neisseria gonorrhoeae*; therefore appropriate antibiotic therapy to cover both of these organisms, such as a third-generation cephalosporin (ceftriaxone) and an oral course of doxycycline, are needed. Other drugs for gonorrhea include cefixime, ciprofloxacin, ofloxacin, or levofloxacin. Chlamydia infection also responds to azithromycin and ofloxacin.

5 Gram-Positive Cocci

Q: A 67-year-old man was hospitalized because of complications from influenza. The patient then developed a bacterial pneumonia that was life threatening. The bacteria in the lung spread to the blood. Blood culture resulted in the isolation of gram-positive cocci that were catalase and coagulase positive. What is the most likely pathogen?

    A: *Staphylococcus aureus*. All staphylococci are catalase positive, whereas all streptococci are catalase negative. *S. aureus* is coagulase positive, whereas other staphylococci are coagulase negative.

Q: What is the most common cause of bacterial pharyngitis?

    A: *Streptococcus pyogenes* (group A streptococcus). Be sure to learn the name of the organism, not just the group designation.

Q: What is the name of the grouping scheme for the streptococci and on what bacterial characteristic is it based?

    A: Many of the beta-hemolytic streptococci are serotyped and placed into various Lancefield groups (after Rebecca Lancefield). The C-carbohydrate in the cell wall is the basis for the group-specific antigens. Not all streptococci can be placed into Lancefield groups (e.g., viridans streptococci).

Q: A 72-year-old patient with an artificial hip cut her head during a fall in her home. Because of a serious concussion, she was hospitalized. Within a few days, she developed a high fever. Gram-positive, catalase-positive, coagulase-negative cocci were isolated from a blood culture. The focal point of infection was subsequently shown to be the surface of the artificial hip. What is the most likely organism?

    A: *Staphylococcus epidermidis* (gram-positive cocci = staph, strep, or enterococci; catalase positive = staph; coagulase negative = *S. epidermidis*, *Staphylococcus saprophyticus*, others; surface of artificial hip = *S. epidermidis*)

Q: A 68-year-old man presented with sudden onset of headache, high fever, vomiting, and stiff neck. Meningitis was suspected, and cerebrospinal fluid was drawn and sent to the diagnostic lab. A Gram stain of spun cerebrospinal fluid revealed gram-positive diplococci. What is the empiric antimicrobial therapy?

    A: The organism is most likely *Streptococcus pneumoniae* (gram-positive diplococci). Although therapies change over time, for *S. pneumoniae* meningitis vancomycin plus ceftriaxone or cefoxitin is one therapeutic approach. Susceptibility testing should be done because of changing resistance patterns.

Q: What organism is responsible for scarlet fever?

    A: *Streptococcus pyogenes* (group A streptococcus)

Q: What virulence factor of *S. pyogenes* is responsible for the rash of scarlet fever?

    A: Pyrogenic toxin (formerly erythrogenic toxin). There are four immunologically distinct types, namely A, B, C, and F. These serotypes are not related to the Lancefield serotyping of the streptococci (groups A, B, etc.).

Q: What is the mode of action of the pyrogenic (erythrogenic) toxins in scarlet fever?

    A: They are superantigens and, like all superantigens, can polyclonally activate large fractions of the T-cell population at picomolar concentrations. These toxins are also responsible for streptococcal toxic shock syndrome and can result in systemic shock and death. Other superantigens include the staphylococcal toxic shock syndrome toxin-1 (TSST-1). Recent evidence suggests that the rash of scarlet fever may be due to a hypersensitivity reaction rather than as a direct effect of the toxin.

Q: What is the most common cause of acute rhinosinusitis?

    A: Viruses are the most common causes, but the condition can be complicated with bacteria, especially *Streptococcus pneumoniae*. *Haemophilus influenzae* and *Moraxella catarrhalis* (both of the latter are gram-negative) are also common.

Q: What is streptolysin O and why is it important?

A: Streptolysin O is a hemolysin produced by *Streptococcus pyogenes*. Antibodies to streptolysin O (ASO) are important in diagnosing post-streptococcal sequelae, such as rheumatic fever. High ASO titers suggest a prior *S. pyogenes* infection in suspected cases of rheumatic fever.

Q: What bacterium causes scalded skin syndrome?

A: *Staphylococcus aureus*

Q: What are the major characteristics used to presumptively identify *S. aureus*?

A: Gram-positive cocci in clusters, catalase and coagulase positive

Q: What test is used to differentiate between the staphylococci and the streptococci?

A: Catalase test. All staphylococci are catalase positive, whereas all streptococci (including *Enterococcus*) are catalase negative. Catalase is not unique to the genus *Staphylococcus*. There are other cocci (such as *Micrococcus*) and many bacilli (most aerobes, facultative anaerobes, and even a few anaerobes) that are catalase positive. The bottom line is that most bacteria are catalase positive, so just learn those that are negative (strep and enterococci are most important; do not worry about the rest).

Q: How is the catalase test performed?

A: The bacteria from a colony are suspended in a solution of hydrogen peroxide (e.g., such as on a microscope slide). Catalase-positive bacteria cause the breakdown of hydrogen peroxide into water and oxygen. The oxygen is evident by the formation of gas bubbles (positive test).

Q: What test is used to differentiate between *Staphylococcus aureus* and all other staphylococci?

A: Coagulase test. *S. aureus* is coagulase positive, and all others are negative. This is all you need to know about how to identify species of the staphylococci.

Q: How is the coagulase test performed?

A: The bacteria from a colony are suspended in a solution of rabbit plasma. After incubation (30 minutes to 24 hours), coagulase-positive bacteria cause the plasma to solidify or gel.

Q: What species is responsible for staphylococcal food poisoning?

A: *Staphylococcus aureus*

Q: After determining that bacteria are either *Streptococcus* or *Enterococcus* (using the Gram stain and the catalase test), the next step in identification requires looking at the hemolytic patterns on blood agar plates. What are the names of the three hemolytic patterns and what do they look like?

A: Nonhemolytic (gamma hemolysis; blood remains red, no clearing or color change), alpha-hemolysis (partial or incomplete hemolysis that turns blood in plate a greenish color), and beta-hemolysis (complete hemolysis that causes all the red blood cells to lyse, revealing the normal clear yellow coloration of the agar medium)

Q: Which major species or groups of streptococci are alpha-hemolytic and which are beta-hemolytic?

A: Alpha: *Streptococcus pneumoniae* (pneumococcus) and the viridans group (there is no organism called *Streptococcus viridans*). Beta: *Streptococcus pyogenes* (this is group A, but learn the species name for standardized exams) and *Streptococcus agalactiae* (group B, but learn the species name). These are the only hemolytic patterns you need to learn for the streptococci. All other species, including *Enterococcus*, are either nonhemolytic or have variable patterns that you do not need to learn. *Enterococcus* can be differentiated from the *Streptococcus* by other tests.

Trivia: There are some rare variants of *S. pyogenes* and *S. agalactiae* that are nonhemolytic. You do not need to know this for any standardized exams. As an added note, I tell my own students that there are exceptions to almost everything I tell them, but typically they are not responsible for knowing these.

Q: Because both *Streptococcus pneumoniae* and the viridans group are alpha-hemolytic, how can you tell them apart?

A: The Gram stain helps. *S. pneumoniae* are gram-positive lancet-shaped diplococci (you might not see the lancet shape, but you cannot miss the diplococci). The viridans group streptococci are typically longer chains of gram-positive cocci. Other tests include the optochin test (*S. pneumoniae* sensitive, viridans resistant) and the bile solubility test (*S. pneumoniae* soluble, viridans group not soluble). The quellung test can also be used to identify *S. pneumoniae*.

Q: What is the Quellung test?

A: The Quellung test is used to either identify or serotype encapsulated species of bacteria, especially *S. pneumoniae*. It utilizes antisera prepared against the polysaccharide capsule. In the presence of specific antisera, the capsule will appear to swell, or become more refractile, making the cells easier to visualize. The test is done on a microscope slide. After the antiserum is added, the bacteria are viewed microscopically. Identification uses polyvalent antiserum (many or all serotypes) and typing uses monovalent serum (against one serotype). As mentioned earlier, knowledge of the quellung reaction is not high yield.

Q: Describe the typical clinical presentation of pneumococcal pneumonia.

A: It can present as a lobar pneumonia or bronchopneumonia, having an acute onset with shaking chills, high temperature, and chest pain. A productive cough with rust-colored (bloody) sputum is common. Acute onset of chills, fever, and chest pain are typical symptoms of bacterial pneumonias and are not specific for *Streptococcus pneumoniae*.

Q: What is the major virulence factor of *S. pneumoniae*?

A: Primarily it is the antiphagocytic polysaccharide capsule.

Q: Name some species of viridans group streptococci. Where are they normally found?

A: *Streptococcus mutans*, *S. salivarius*, *S. mitis*, and *S. sanguis* are some common ones. They are found primarily in the oral cavity but can be found elsewhere in the gastrointestinal tract.

Q: Why are the viridans streptococci important?

A: They can cause subacute bacterial endocarditis. Some, such as *S. mutans*, contribute to tooth decay (dental caries).

Q: *Streptococcus pyogenes* and *Streptococcus agalactiae* are both beta-hemolytic. What laboratory tests will help you tell them apart?

A: *S. pyogenes* is sensitive to bacitracin. *S. agalactiae* is resistant. In addition, *S. agalactiae* is positive in the CAMP test, and it can hydrolyze hippurate.

Q: What is the CAMP test?

A: It is performed on a blood agar plate using a known laboratory strain of beta-hemolytic *Staphylococcus aureus* and an unknown patient isolate of *Streptococcus*. The *S. aureus* is streaked down the center of the blood agar plate. The unknown *Streptococcus* is streaked at right angles to the *S. aureus* so that the two streaks touch. After overnight incubation, the plates are observed for an area of enhanced beta-hemolysis where the two streaks make right angles to each other. If enhanced beta-hemolysis is observed, it is a positive CAMP test and the *Streptococcus* is probably *S. agalactiae*.

Trivia: CAMP is an acronym for Christie, Atkins, Munch, and Petersen, the discoverers of the phenomenon.

Q: *Streptococcus pyogenes* causes what kinds of infections?

A: Pharyngitis, impetigo (pyoderma), erysipelas, cellulitis, necrotizing fasciitis, and toxic shock syndrome are some of the most notable. As with many other bacteria, bacteremia can develop, particularly from cellulitis and necrotizing fasciitis.

Q: What are the major virulence factors of *S. pyogenes* and how do they function in streptococcal disease?

A: The M protein, also called fimbriae, is a component of the cell wall and is antiphagocytic and functions in adherence. There are several kinds of pyrogenic (erythrogenic) exotoxins that function as superantigens and are responsible for the rash of scarlet fever and streptococcal toxic shock syndrome. Streptolysin O probably functions as a virulence factor in some diseases of streptococcal infections, but it is more important for being antigenic and inducing antibodies against itself. These antibodies are called antistreptolysin O (ASO) antibodies. Elevated ASO titers are indicative of recent *S. pyogenes* infection and can be useful in diagnosis of poststreptococcal sequelae (e. g., scarlet fever, rheumatic fever, and glomerulonephritis). Hyaluronidase (spreading factor) is involved in infections such as impetigo. Other virulence factors include streptokinase (fibrin lysis), streptolysin S (an oxygen-stable hemolysin), DNase (degrades DNA), and C5a peptidase (degrades complement component C5a).

Q: *Streptococcus agalactiae* causes what kinds of infections?

A: Primarily neonatal sepsis such as meningitis, pneumonia, and bacteremia, but it can also cause infections in adults. In neonates, there is both an early-onset (usually within the first week of birth) and a late-onset disease.

Q: How does early onset differ in its epidemiology and presentation from late-onset neonatal *S. agalactiae* disease?

A: Early onset is acquired during birth, commonly due to aspiration of contaminated amniotic fluid. It frequently presents with pneumonia, bacteremia, and meningitis. Late-onset disease is by direct contact with a source, such as the mother or a nurse, and does not typically include lung involvement.

Q: What are the recommended guidelines for reducing the incidence of group B streptococcal disease in neonates?

A: There are about a dozen, but the most important for you to remember include (1) screening all pregnant women at 35–37 weeks gestation for vaginal and rectal group B streptococcal colonization and (2) beginning intrapartum chemoprophylaxis for group B streptococcal carriers at the time of labor or rupture of membranes.

Q: With respect to the previous question, what is the drug of choice for intrapartum chemoprophylaxis?

A: Penicillin, but ampicillin is a good alternative. Do not forget to assess allergies of the patient.

Q: What role do the enterococci play in disease?

A: The three primary infections are urinary tract infections (usually nosocomial and frequently catheter associated), bacteremia, and subacute bacterial endocarditis. They can also be involved in other infections such as hepatobiliary sepsis, surgical wound infections, and neonatal sepsis, all of which can lead to bacteremia. There can be an overlap in the time of onset of acute and subacute endocarditis, so be careful in the use of the terms. Although both terms are commonly used, there is a movement away from their usage. Infectious endocarditis is the preferred term, because other organisms (e.g., fungi and viruses) can be a cause.

Q: Name the two major species of enterococci.

A: *Enterococcus faecalis* and *E. faecium*. It is not important to know the differences between these two species. Both are involved in the same kinds of diseases, although *E. faecalis* is more common.

Q: How are the enterococci differentiated from *Streptococcus*?

A: You cannot differentiate based on the Gram stain or hemolytic pattern (usually alpha, but can be nonhemolytic or [rarely] beta), so it must be done biochemically. Common tests include the ability of *Enterococcus* to grow in the presence of bile, to grow in 6.5% salt, and to hydrolyze esculin (a carbohydrate). These are the most important tests, but there are others (e.g., growth at 45°C and hydrolysis of pyrrolidonyl-beta-naphthylamide; do not memorize this!). Hydrolysis of esculin and growth in bile can be done on the same growth medium. If it turns black, the organism grew and hydrolyzed the esculin. Be careful of the term bile positive. It can mean growth in the presence of bile (*Enterococcus*) or bile soluble (*Streptococcus pneumoniae*). It would be worthwhile for you to use the information presented thus far in this section to draw an identification scheme for the gram-positive cocci.

Q: What is the drug of choice for enterococcal infections?

A: Choose penicillin (or another beta-lactam) if given a choice. More serious infection might need the addition of an aminoglycoside, but this is unlikely to be asked.

Q: *Streptococcus bovis* isolated from a bacteremic patient is indicative of what?

A: Possible cancer of the colon. This is all you need to know about this organism.

Q: In a discussion about possible causes of intra-abdominal abscesses, someone mentioned "anaerobic strep." What genus were they most probably referring to?

A: *Peptostreptococcus*. The term "anaerobic streptococcus" is jargon.

Q: What is the composition of the two primary pneumococcal vaccines?

A: The vaccine recommended for those ages 2 or older is composed of the purified capsular polysaccharide antigens of 23 (23 valent) of the most common serotypes (there are about 90 serotypes). This covers nearly 90% of all *S. pneumoniae* infections. The pediatric pneumococcal conjugate vaccine contains the purified capsular polysaccharide antigens of 7 serotypes (7 or hepta-valent). It can be used for children under the age of 2.

Trivia: The polysaccharide antigens are conjugated to a nontoxic mutant diphtheria toxin to improve their antigenicity in children.

Q: Name a bacteriological medium that will select for the isolation and growth of *Staphylococcus aureus* but inhibit the growth of most other bacteria. What component of the medium inhibits the growth of most bacteria?

A: Mannitol salt agar contains 7.5% sodium chloride. *S. aureus* can grow at this salt concentration, but most other bacteria cannot.

Q: What is the natural habitat of *S. aureus*?

A: Primarily normal flora of the skin, including the nares, but it is also part of the intestinal microflora. In addition, it is found as normal flora of many animals.

Q: What is the hemolytic pattern of *S. aureus*?

A: Most strains are beta-hemolytic, but you usually do not need to know this for answering questions about *S. aureus*. It is more important to know they are catalase and coagulase positive.

Q: What are some virulence factors of *S. aureus*?

A: Protein A, Panton-Valentine leukocidin, teichoic acids, alpha toxin, enterotoxins, exfoliatin, and TSST-1 are some of the most important. Others include various enzymes such as hemolysins, leukocidin, hyaluronidase, lipase, coagulase, and fibrinolysin and toxins such as sphingomyelinase. Not all strains have all these factors. The possession of specific virulence factors is genetically determined and helps determine the specific diseases various strains are capable of causing.

Q: What are the modes of action of the primary *S. aureus* virulence factors listed in the previous question and in what diseases are they important?

A: Protein A is a structural component of the cell wall. It has antiphagocytic properties (it blocks opsonization) by binding to the Fc region of IgG. The Panton-Valentine leukocidin is a pore-forming toxin secreted by less than 5% of virulent strains of *S. aureus*. These strains are capable of causing severe soft tissue infections and necrotizing pneumonia. Teichoic acids are structural components of the cell wall that function in attachment (colonization) of the bacteria to fibronectin of mucosal cells. Alpha toxin causes membrane damage and is partly responsible for the bacteria spreading into tissues. All three of these virulence factors are important in most infections, especially abscesses. There are five serologically distinct enterotoxins that can be involved in food poisoning and related disease. The enterotoxins are multifunctional: They have superantigen properties (cytokine induction) and neurotoxin-like activity (affect the vomiting center of the central nervous system). They are also responsible for the cramping, diarrhea, and nausea of staphylococcal food poisoning. You do not need to know the specific mechanism of exfoliatin. Just know that it is responsible for the erythema and bullae of staphylococcal scalded skin syndrome. TSST-1, the only staphylococcal toxic shock syndrome toxin so far, is a superantigen that induces the cytokines responsible for the disease symptoms. The mechanism of action of superantigens is similar to that of endotoxin.

Q: What is the route of transfer between hospitalized patients of multiresistant *Staphylococcus aureus*?

A: Indirect transfer, probably via the hands, from one patient to a health care provider (doctor, nurse, phlebotomist, etc.) and then to another patient

Q: With respect to the previous question, what is the best way to minimize the spread of pathogens from one patient to another?

A: Hand-washing between visiting different patients

Q: What is the most common cause of community-acquired pneumonia?

A: *Streptococcus pneumoniae*. Other common causes of community-acquired pneumonia include *Haemophilus influenzae*, *Moraxella catarrhalis*, *Staphylococcus aureus*, and *Legionella pneumoniae*. *Chlamydophila pneumoniae* and *Mycoplasma pneumoniae* are also common (walking pneumonia), especially in younger patients.

Q: What is the drug of choice for community-acquired pneumonia of unknown etiology in an adult with no other comorbidities?

A: A macrolide or doxycycline is a good choice (these also are effective against *Mycoplasma pneumoniae* and *Chlamydophila pneumoniae*). A fluoroquinolone or a cephalosporin is also good as empirical treatment. Therapy can be modified after susceptibility results become available.

Q: What is the drug of choice for pneumococcal pneumonia?

A: If the isolate is susceptible, penicillin G is the drug of choice. However, penicillin resistance can be as high as 35%. In resistant strains, a fluoroquinolone (e.g., levofloxacin, gatifloxacin, or moxifloxacin) or doxycycline is a good choice.

Q: What is the basis for resistance in methicillin-resistant *Staphylococcus aureus* (MRSA; i.e., what is the molecular basis for resistance)?

A: Resistance is due an alteration in a penicillin binding protein that confers low affinity for the methicillin. Therefore, the methicillin can no longer bind to the penicillin binding protein.

Q: What is the normal function of penicillin binding proteins?

A: They are enzymes (transpeptidases) that function in the final steps of peptidoglycan (cell wall) synthesis.

Q: Are MRSA resistant to other antibiotics besides methicillin?

A: Yes. Methicillin is one of several beta-lactamase–resistant penicillins. Others are nafcillin, dicloxacillin, cloxacillin, oxacillin, and flucloxacillin. Not all are readily available in the United States. If *Staphylococcus aureus* is resistant to methicillin, it is also usually resistant to the others as well. Therefore MRSA is used as a general term for any *S. aureus* strains that are resistant to any of the beta-lactamase–resistant penicillins. MRSA are commonly resistant to other non-beta-lactam antibiotics and therefore can be difficult to treat.

Q: What is the drug of choice for MRSA?

A: Vancomycin (+/– gentamicin +/– rifampin). The vancomycin is most important to remember.

Q: What is the disease association of *Staphylococcus saprophyticus*?

A: Urinary tract infections in sexually active young women are most common.

Q: What is the disease association of *Staphylococcus epidermidis*?

A: Infections are commonly (but not exclusively) associated with the use of artificial implants of any kind, including artificial heart valves, shunts, and joints, as well as catheters. It is important to remember the catheter association.

Q: What is the primary virulence factor of *Staphylococcus epidermidis*?

A: *S. epidermidis* possesses a capsule (usually called a slime layer because it is more descriptive) that allows the organism to stick tenaciously to artificial surfaces. After it sticks, it reproduces to form a biofilm that is difficult for the immune system or antimicrobials to penetrate. Treatment frequently includes removal of the catheter or implant and the use of antibiotics.

Q: What are the primary symptoms of staphylococcal food poisoning?

A: Rapid onset (4–6 hours) with vomiting, usually with abdominal cramping. Diarrhea may also be present. Symptoms usually subside within a day.

Q: Why does staphylococcal food poisoning have a rapid onset?

A: It is an intoxication. The bacteria grow to high numbers in the food where they secrete the enterotoxin that is then ingested. Because the toxin is preformed, there is a short lag time between ingestion and symptoms.

Q: What kinds of foods are typically associated with staphylococcal food poisoning?

A: Nonrefrigerated foods that are high in protein (e.g., ham and other meats, gravies, turkey dressing, chicken salad, custard-filled pastries, etc.)

Q: What effect does reheating the contaminated food have on *Staphylococcus aureus* and its enterotoxin?

A: It kills the bacteria but has little effect on the heat-stable toxin.

Q: What is the therapy for staphylococcal food poisoning?

A: Usually none. Rehydration may be necessary, especially in young children or older adults.

Q: What is the most common cause of otitis media?

A: *Streptococcus pneumoniae* (~50% of all cases). Two other common causes are *Haemophilus influenzae* (~25%) and *Moraxella catarrhalis* (~10%).

Q: In broad terms, what is the age distribution for otitis media?

A: It is primarily a childhood disease in those younger than 3 years old.

Q: What is the treatment for otitis media?

A: It depends on the age and the severity of the infection. In general terms, it should be treated with antibiotics if the child is very young (<6 months). In older children, an observation period of up to 3 days might be recommended. In any case, pain management should be part of the treatment.

Q: What is the drug of choice for acute otitis media?

A: Primarily, oral amoxicillin. Second choices include amoxicillin + clavulanate or oral cefuroxime axetil. Drug hypersensitivities should be considered.

Q: What causes toxic shock syndrome?

A: The most common cause of toxic shock syndrome is *Staphylococcus aureus*, associated either with menstruation or with a focal infection. Menstruation-associated cases have steadily decreased since the early 1980s. *Streptococcus pyogenes* is a less common cause, with most cases due to a focal infection other than strep throat.

Q: What are some signs and symptoms of toxic shock syndrome?

A: High fever, myalgia, diarrhea, and vomiting early in the disease; diffuse macular rash; followed by desquamation (especially the palms and soles); and hypotension. It can lead to multiorgan failure and death.

Q: What is the most common cause of osteomyelitis?

A: *Staphylococcus aureus*

Q: Bullous impetigo is related to what other staphylococcal disease?

A: Bullous impetigo is a highly infectious localized form of staphylococcal scalded skin syndrome. As with staphylococcal scalded skin syndrome, bullous impetigo is most common in children and infants.

Q: What is the most common cause of prosthetic heart valve endocarditis?

A: *Staphylococcus epidermidis*

Q: What is the drug of choice for an infection due to *S. epidermidis*?

A: It is the same as for *Staphylococcus aureus*, that is, a penicillinase-resistant penicillin. If the strain is methicillin resistant (resistant to penicillinase-resistant penicillins), then vancomycin is the drug of choice.

Q: *Streptococcus pyogenes* can cause soft tissue infections. What is the "spreading factor" that allows *S. pyogenes* to spread into healthy tissue?

A: Hyaluronidase

Q: A 72-year-old febrile male patient presented with acute onset of spreading erythema and edema on his right leg. The involved skin was warm and slightly raised, and the edge was sharply demarcated from uninvolved skin. What is the catalase reaction of the organism involved?

A: Catalase negative. This is most likely a case of erysipelas and is most commonly due to *Streptococcus pyogenes*.

Q: A 4-year-old boy developed several small lesions on the face. As the lesions healed, the serous exudate dried to give a "honey-crust" appearance. What serious complication can occur with this infection?

A: Acute glomerulonephritis can occur in rare cases.

Q: From the previous question, what is the most likely etiological agent?

A: Most likely *Streptococcus pyogenes*. *Staphylococcus aureus* impetigo can occasionally resemble streptococcal impetigo, but without the glomerulonephritis.

Q: What is the mechanism of the acute glomerulonephritis that is associated with streptococcal impetigo?

A: It is due to the deposition of immune complexes in the kidneys.

Q: What is Ritter disease?

A: Staphylococcal scalded skin syndrome

Q: Vancomycin-resistant enterococci can be successfully treated with what combination of antibiotics?

A: Linezolid and quinupristin/dalfopristin. Linezolid is an oral oxazolidinone with activity against *Enterococcus faecium* and *E. faecalis*. It is also active against MRSA and penicillin-resistant *Streptococcus pneumoniae*. Quinupristin/dalfopristin is indicated for bacteremia due to vancomycin-resistant *E. faecium*, but not *E. faecalis*. It can also be used in skin infections due to methicillin-susceptible *Staphylococcus aureus* or *Streptococcus pyogenes*.

Gram-Negative Cocci

Q: Name the most important genera of gram-negative cocci.

A: *Neisseria* and *Moraxella*. There are others, such as *Veillonella* (an obligate anaerobe), but you do not need to know its shape or Gram stain on most exams. Do not confuse the gram-negative coccobacilli (e.g., *Haemophilus*, *Bordetella*, *Eikenella*, *Kingella*, others), which are true but very short bacilli, with the gram-negative cocci.

Q: How are *Neisseria* classified based on oxygen utilization?

A: They are aerobes.

Q: Are *Neisseria* oxidase positive or negative?

A: Oxidase positive. It is safe for you to assume that all aerobes are catalase positive. In fact, it is usually safe to assume that if the organism in question is not in the family Enterobacteriaceae (*Escherichia coli*, *Salmonella*, *Shigella*, *Klebsiella*, *Proteus*, etc.), it is probably oxidase positive. There are many exceptions to these rules, but you do not need to know them unless you work/run a microbiology lab or specialize in infectious diseases.

Q: Describe the microscopic appearance of *Neisseria*.

A: Gram-negative diplococci that are "kidney bean" shaped with their flat sides toward each other

Q: Name the two most important species of *Neisseria*.

A: *Neisseria gonorrhoeae* and *Neisseria meningitidis*. There are actually about 10 more, but most of you will never see them in human infections.

Q: How can you differentiate between *N. gonorrhoeae* and *N. meningitidis* based on sugar utilization patterns?

A: *N. gonorrhoeae* uses only glucose and *N. meningitidis* uses maltose and glucose. You will often see this discussed as sugar fermentation patterns, but *Neisseria* are not fermentative. They are aerobes; therefore they oxidize these sugars to produce acid oxidatively.

Q: What is the Waterhouse-Friderichsen syndrome?

A: Primary adrenal insufficiency due to adrenal hemorrhage. It is often secondary to meningococcemia-(*N. meningitidis*) induced shock.

Q: What virulence factor of *N. gonorrhoeae* allows it to adhere to mucosal surfaces?

A: The pili are the primary adhesins.

Q: How does phase, or antigenic, variation of the pili of *N. gonorrhoeae* function as a virulence factor?

A: It allows the organism to evade immune surveillance, which may contribute to repeated infections in the same person.

Q: *N. gonorrhoeae* secretes a protease that is an important virulence factor in that it aids in colonization of mucous membranes. How?

A: It is a protease that degrades IgA at mucosal surfaces.

Q: How do neonates normally get infected with *N. gonorrhoeae*?

A: Infection occurs by direct contact with the birth canal during birth.

Q: How does infection of neonates with *N. gonorrhoeae* normally manifest itself?

A: It usually occurs as an eye infection (ophthalmia neonatorum). It is generally thought that in North America *Chlamydia* is about two times more common (about 6 per 1000 live births) as a cause of ophthalmia neonatorum than is *Neisseria*. Infection with *N. gonorrhoeae* usually is more severe and has more complications.

Q: How is ophthalmia neonatorum prevented?

A: Several prophylactic measures can be used, including topical silver nitrate, erythromycin, or tetracycline. Povidone-iodine ophthalmic solution has also been shown to be effective.

Q: How does infection with *N. gonorrhoeae* manifest itself *most often* in women?

A: Most women are asymptomatic. Symptomatic women can present with a purulent vaginal discharge, fever, dysuria, frequency of urination, and red tender mucosa.

Q: How does infection with *N. gonorrhoeae* manifest itself *most often* in men?

A: Men are usually symptomatic, but the incubation period can be longer than a month. A purulent yellow-green urethral discharge, dysuria, and frequency are common. The severity of symptoms is variable. Bacteremia can result in septic arthritis. *N. gonorrhoeae* is the number one cause of arthritis (usually in the knees) in the 18–30-year-old age group.

Q: How is the diagnosis of gonorrhea made in asymptomatic women?

A: Culture and definitive identification. Culture is also required for definitive diagnosis in asymptomatic men and is more reliable than microscopy in symptomatic women. *N. gonorrhoeae* can be identified by a variety of tests, including biochemical and molecular methods. Culture offers the advantage of doing susceptibility testing.

Q: Why is microscopy more reliable in symptomatic men than in symptomatic women for diagnosis of gonorrhea?

A: Men do not generally have gram-negative diplococci as normal flora of the urethra; therefore the observation of gram-negative diplococci in a purulent urethral discharge, particularly within polymorphonuclear neutrophilic leukocytes (PMNs), is diagnostic. In women, the microscopic observation of gram-negative diplococci may indicate normal flora. The test becomes more reliable if gram-negative diplococci are seen within neutrophils, but generally it is best to confirm with culture.

Q: What media should be used for the culture of *N. gonorrhoeae*?

A: Both selective and nonselective media should be used. A primary selective medium is modified Thayer-Martin, but do not worry about the "modified" part of the name. Thayer-Martin medium is a chocolate agar base containing antibiotics (vancomycin, colistin, and nystatin) that suppress the normal flora but usually allow for the growth of *N. gonorrhoeae*. Some strains of *N. gonorrhoeae* do not grow on Thayer-Martin. For this reason, nonselective media such as chocolate agar should be inoculated as a backup.

Q: What is chocolate agar?

A: Chocolate agar is prepared by heating sheep blood (usually) to about 80°C before it is added to the melted agar base. This lyses the red blood cells, releasing the intracellular nutrients that nonhemolytic bacteria have a difficult time obtaining.

Q: A 20-year-old man walked into the emergency room with symptoms highly suggestive of gonorrhea. A swab of the urethral discharge was placed into transport media and refrigerated until it could be sent to the microbiology lab for culture. After delivery to the lab four hours later, the swab was streaked onto appropriate media. After 48 hours incubation, the lab reported only normal flora. Assuming he did indeed have gonorrhea, why did the laboratory fail to isolate *N. gonorrhoeae*?

A: The organism is relatively fragile to drying and cold temperatures, so it was probably dead before it was inoculated onto the media. Specimens should be inoculated onto media (preferably prewarmed) immediately after collection and incubated (preferably immediately). If a delay in culturing cannot be avoided, the specimen should be placed into approved transport media and held at room temperature. Note: Rayon or Dacron collection swabs are preferred over calcium alginate or cotton (the latter may contain toxic fatty acids).

Q: *N. gonorrhoeae* infections in females can cause what diseases?

A: Besides gonorrhea, it can cause infections such as cervicitis, salpingitis, pelvic inflammatory disease, and endometritis. Severe infections can lead to sterility. Arthritis (bacteremic spread) is more common in females than in males.

Q: How do patients get pharyngitis due to *N. gonorrhoeae*?

A: Oral sex. The infection is usually asymptomatic in both females and males.

Q: What is the drug of choice for gonorrhea?

A: The antibiotic that first comes to mind is ceftriaxone, but cefixime, ciprofloxacin, gatifloxacin, and ofloxacin are also good choices. Patients are also usually treated with azithromycin or doxycycline for (presumed) chlamydial infection.

Q: Many gram-negative bacteria have lipopolysaccharide (LPS) as a component of their cell walls. The bacteria in the genus *Neisseria* (among others, e.g., *Haemophilus, Campylobacter*), have lipo-oligosaccharide (LOS). What is the difference between LPS and LOS?

    A: LOS is analogous to LPS. LOS, however, does not have the O antigens found in LPS. For standardized exams, you do not need to know the difference. Just recognize that LOS is the endotoxin of the *Neisseria* and a few other gram-negative bacteria.

Q: What is the natural reservoir of *N. meningitidis* and *N. gonorrhoeae*?

    A: Humans. They are not found elsewhere.

Q: What age group is at highest risk for *N. meningitidis* infection?

    A: Generally, about 3–15 years of age, but especially children under 5. People at all ages are at risk. It is the second most common cause of bacterial meningitis in adults (*Streptococcus pneumoniae* is number one). Do not memorize the exact ages because it is listed differently in various references.

Q: What are the symptoms and signs of meningococcal meningitis?

    A: They are the same as any bacterial meningitis in children and adults: abrupt onset of fever, chills, headache, stiff neck, vomiting, and irritability. Brudzinski and Kernig signs are usually present.

Q: List the primary virulence factors of *N. meningitidis*.

    A: Antiphagocytic polysaccharide capsule, IgA protease, endotoxin (LOS), and pili are the ones you should know. Others include several outer membrane proteins that function in iron acquisition or apoptosis of human cells. *Neisseria* do not secrete siderophores. *N. gonorrhoeae* also has a polysaccharide capsule, but it does not seem to be as important as a virulence factor as it is for *N. meningitidis*.

Q: What is a cutaneous manifestation seen in *N. meningitidis* bacteremia (meningococcemia)?

    A: A petechial skin rash that may coalesce to form hemorrhagic bullae

Q: Of the virulence factors listed above, which one is responsible for the skin rash seen in meningococcemia?

    A: LOS, which is the endotoxin

Q: How is the diagnosis of meningitis due to *N. meningitidis* made?

    A: Presumptive diagnosis is made either by direct observation by Gram stain of the cerebrospinal fluid (CSF; after centrifugation at 2000 rpm for 20 minutes) or by a latex agglutination test (detection of specific bacterial capsular antigens in the CSF). Definitive diagnosis is based on culture of the organism.

Q: What is the microscopic appearance of *N. meningitidis* in a positive CSF specimen and how can it be differentiated from *Streptococcus pneumoniae* and *Haemophilus influenzae*?

    A: Gram-negative kidney (or coffee) bean–shaped diplococci either within PMNs or extracellularly (the more PMNs in the CSF, the better the prognosis; therefore quantitation of bacteria and PMNs should be reported). *Streptococcus pneumoniae* are gram-positive lancet-shaped diplococci, and *Haemophilus influenzae* are small gram-negative coccobacilli or pleomorphic rods (often times elongated).

Q: Untreated (or delayed treatment) of *N. meningitidis* meningitis can result in a death rate approaching 100%. What is the prognosis in those who are promptly treated?

    A: Much better, with a mortality rate less than 10%

Q: What is the drug of choice for *N. meningitidis* meningitis?

    A: Penicillin G

Q: How can *N. meningitidis* infections be prevented?

    A: Vaccination

Q: What is the composition of the meningococcal vaccine?

    A: It is composed of the organism's polysaccharide capsule. The U.S. Food and Drug Administration has licensed a polysaccharide vaccine for serogroups A, C, Y, and W-135. A single dose of vaccine is all that is required. It is recommended that persons at risk, such as college students living in dormitories, receive the vaccine.

Q: If an outbreak of *N. meningitidis* infection occurs in an area of close quarters (e.g., military barracks, dormitory, etc.), what prophylactic antimicrobials should be used?

    A: Rifampin is the first choice, but there are some resistant strains. Other drugs include ciprofloxacin, ceftriaxone, and azithromycin. Susceptibility testing of the primary isolate can be useful in making the decision.

Q: What is the clinical significance of *Moraxella catarrhalis*?

A: *M. catarrhalis* is a gram-negative diplococcus resembling the *Neisseria* morphologically. It was formerly called *Neisseria catarrhalis* and *Branhamella catarrhalis*. It is the third most important cause (after *Streptococcus pneumoniae* and *Haemophilus influenzae*) of otitis media and rhinosinusitis in children. It is a minor cause of bronchitis and is an occasional cause of other infections such as septicemia, meningitis, endocarditis, conjunctivitis, laryngitis, and pneumonia. Elderly patients, patients with other pulmonary disease, and immunocompromised patients are at high risk.

Gram-Positive and Acid-Fast Bacilli

Q: What are the three forms of anthrax and which one is most common?
> A: The most common form is cutaneous anthrax. The other two forms are gastrointestinal (rarest form) and pulmonary.

Q: What is the natural reservoir of *Bacillus anthracis*?
> A: Herbivores

Q: A 47-year-old alcoholic had a change in health status that included fever and chills, loss of appetite and weight, night sweats, and fatigue. A Gram stain of sputum was unrevealing, but an acid-fast stain demonstrated small red bacilli. What is the diagnosis?
> A: Pulmonary tuberculosis (TB) due to *Mycobacterium tuberculosis*. Night sweats are not unique to TB. It occurs in other diseases such as AIDS (most likely due to *M. avium*), brucellosis, pulmonary histoplasmosis (must be differentiated from TB), subacute endocarditis, and chronic pneumonia.

Q: The above patient has *active* pulmonary TB. What is the difference between active TB and latent TB?
> A: In active TB, the patient has signs and symptoms of TB disease. In latent TB, the patient is purified protein derivative (PPD) positive (he is infected) but is asymptomatic.

Q: Briefly, describe the pathophysiology leading to latent or *chronic* pulmonary TB. Note: this is best done by drawing the steps on a piece of paper.
> A: Primary infection begins with inhalation of droplet nuclei into the lower regions of the upper lobes of the lungs or into the lower lobes. The mycobacteria are phagocytized by alveolar macrophages within which they survive, multiply, and lyse the macrophages. Free bacteria are generally rephagocytized (although sometimes some may disseminate), and the process is repeated. Much of this activity occurs in the hilar lymph nodes. As the macrophages die, small granulomatous lesions (tubercles) form that have a central area of caseous necrosis. Some mycobacteria remain viable in the tubercles. As the cell-mediated immune response continues, the tubercles heal in the form of calcified granulomatous lesions. These calcified lesions, together with the draining hilar lymph nodes, are referred to as the Ghon complex and are visible by radiography. Most patients are asymptomatic with latent (chronic) TB. The patient would be PPD positive.

Q: Briefly describe the pathophysiology leading to active pulmonary TB.
> A: Generally, the disease initially develops as described above for latent TB, but sometimes it can develop into active TB without becoming latent. Risk factors that allow latent disease to progress to active disease (reactivation) include alcoholism, renal failure, malnutrition, or any event that results in a compromised immune system. As mentioned in the previous question, some mycobacteria remain viable in the tubercles. Reactivation results in hematogenous spread including to the apices of the lungs (high oxygen; *M. tuberculosis* is an aerobe), which are primary sites for secondary active TB. Cavitary lesions with caseous necrosis occur in this stage. Because of the erosion of the lungs, hemoptysis (bloody sputum) can be present. The disease can develop into miliary TB and affect many other internal organs.

Q: How is the diagnosis of TB disease made?
> A: Traditionally, diagnosis is based on positive chest x-ray, positive acid-fast stain of sputum, positive PPD skin test, and culture for *M. tuberculosis*. Note that a positive acid-fast stain by itself is not diagnostic because there are other acid-fast organisms. In addition, the absence of observable acid-fast bacilli is common in active cases.

Q: Although new methods such as polymerase chain reaction and other molecular probes, enzyme-linked immunosorbent assay, and interferon-gamma release assays have been or are being developed for diagnosis of TB, most laboratories still grow the organism for positive identification. What medium do they use?

A: Löwenstein-Jensen is one of the most common media. It can take from four to eight weeks for colonies to develop.

Q: What is the interferon-gamma release assay (IGRA)?

A: It is a test that can be used as a tool for the diagnosis of active disease and latent *M. tuberculosis* infection. It measures the amount of interferon-gamma produced by cells in whole blood (from the patient) that have been stimulated by mycobacterial peptides. The U.S. Food and Drug Administration approved and licensed the first available test in 2005. The peptides used in the test are present in *M. tuberculosis* but absent from all bacillus Calmette-Guérin (BCG) strains and from most commonly encountered non-TB mycobacteria. As with the skin test, it cannot distinguish between latent infection and active TB disease. The advantages of the IGRA compared with the skin test are that the results can be obtained after a single patient visit, the variability associated with skin-test reading can be reduced because "reading" is performed in a qualified laboratory, and IGRA is not affected by previous BCG vaccination, thus eliminating the unnecessary treatment of persons with false-positive results.

Q: What are the primary antimicrobial agents used to treat *M. tuberculosis* infections?

A: Until susceptibilities are known, isoniazid, rifampin, pyrazinamide, and ethambutol should be given. Secondary drugs include kanamycin, cycloserine, capreomycin, ethionamide, para-aminosalicylic acid, and several others. Be sure to review the pharmacology of the primary drugs (beyond the scope of this publication).

Q: *M. tuberculosis* strains involved in so-called extensively drug-resistant TB are resistant to what drugs?

A: Isoniazid and rifampin plus resistance to any fluoroquinolone and at least one of three injectable second-line drugs (i.e., amikacin, kanamycin, or capreomycin). As mentioned in an earlier question, it is doubtful that you need to memorize this for standardized exams. You should, however, be aware that resistant strains are becoming common in endemic areas, including areas where HIV infection is high. Resistance to the fluoroquinolones and second-line injectable drugs has been associated with poor treatment outcomes.

Q: What other organisms are acid-fast?

A: Other mycobacteria, *Nocardia*, and *Cryptosporidium* (a parasite)

Q: A 17-year-old foreign exchange student from Russia had been in the United States only four days when she complained of a sore throat, fever, and chills. A throat swab yielded gram-positive pleomorphic rods that contained metachromatic granules. What is the mode of action of the exotoxin that this organism secretes?

A: If you do not know the answer, what are the hints? First, the foreign exchange student in the United States was put in to show that she has a disease not common in the United States (0–2 cases/year in United States; 27–32 /100,000 in Russia and Tajikistan). Second, she has respiratory problems. Third, gram-positive pleomorphic rods were isolated (How many gram-positive pathogens can you even think of? Not too many. How many are pleomorphic? Not many). Finally, the bacteria have metachromatic granules. This is a test to help identify *Corynebacterium diphtheriae*, a gram-positive pleomorphic bacillus that is not a common pathogen in the United States. The exotoxin it secretes inhibits protein synthesis by inactivating elongation factor-2. What you should learn about the toxin is that it causes ADP ribosylation of elongation factor-2. The primary reason to learn about this exotoxin is that it serves as a model of how other exotoxins work, even though their mode of action might be different.

Q: Name another organism that secretes an exotoxin that has the identical mode of action as that of *C. diphtheriae* (Hint: it is gram-negative).

A: Toxin A of *Pseudomonas aeruginosa* has the same mode of action, but it is in no way related to the diphtheria toxin.

Q: What are metachromatic granules?

A: They are stored inorganic phosphate granules that stain red when *C. diphtheriae* is treated with methylene blue.

Q: What is the drug of choice for diphtheria?

A: Erythromycin or penicillin G. These are usually reasonable guesses for gram-positive bacteria if you have no clue.

Q: Can diphtheria be adequately treated with antibiotics only?

A: No. Therapy should include antitoxin. The antitoxin binds to and inactivates any toxin that is still circulating (i.e., unbound toxin).

Q: The 17-year-old foreign exchange student mentioned previously was given antitoxin but went into anaphylactic shock. Why?

A: Because the antitoxin is made in horses and the patient was allergic to horse serum proteins

Q: What is the oxygen requirement for *Corynebacterium diphtheriae*?

A: It is an obligate aerobe.

Q: What kind of bacteriological medium can be used to selectively grow *C. diphtheriae*?

A: Several selective media can be used, but the most common is a tellurite medium (several different formulations). *C. diphtheriae* colonies are black.

Q: What is the diphtheria antigen used in the DTaP and DPT vaccines?

A: Toxoid (inactivated toxin). DPT has largely been replaced in most developed countries with DTaP.

Q: What species of *Mycobacterium* is associated with water and can cause skin infections?

A: Two possibilities here: *Mycobacterium marinum* and *Mycobacterium gordonae*. *M. marinum* is usually saltwater associated but can be found in fresh water. Infections are typically work associated (e.g., fishermen, aquarium workers, etc.). Although a rare pathogen, questions on standardized exams are common (look for a saltwater or aquarium reference). *M. gordonae* is actually quite common in the nonmarine environment (soil, fresh water, whirlpools, mucous membranes of healthy individuals, etc.) but has low pathogenic potential, especially in immunocompetent individuals. Infections can be systemic. It is more common in fresh water. Questions about this organism are not common.

Q: A 37-year-old man in Houston, Texas worked at the shipping docks, primarily preparing imported sheep wool for rail shipment to buyers in other parts of the United States. He noticed a small lesion on his hand that developed into an inflamed pustule that eventually turned black. Based on this information, the most likely infecting organism is what?

A: *Bacillus anthracis*. This is an example of cutaneous anthrax. The pustule is developing into a characteristic eschar. This form of anthrax has a low mortality rate.

Q: List some important characteristics of *B. anthracis* (shape, Gram stain, etc.).

A: Gram-positive, aerobic bacillus, spore former. You should know at least these characteristics.

Q: Pulmonary TB is contagious. Can the same be said for pulmonary anthrax?

A: No. Anthrax is not considered communicable person-to-person.

Q: What radiological finding is suggestive of pulmonary anthrax in a suspected case?

A: Widened mediastinum

Q: What is the approximate mortality rate for untreated pulmonary (inhalational) anthrax?

A: 100%

Q: What are the primary virulence factors of *B. anthracis*?

A: Antiphagocytic capsule and exotoxin. The capsule is composed of polyglutamic acid. There are two exotoxins made up of three components: protective antigen (PA), edema factor (EF, an adenylyl cyclase), and lethal factor (LF, a protease). PA–EF form the edema toxin and PA–LF form the lethal toxin. Note: This description of the toxin composition is oversimplified and will be addressed in another question.

Q: What is the drug of choice for anthrax?

A: Ciprofloxacin or a tetracycline. Most naturally occurring strains are susceptible to penicillin G unless they have been genetically modified to be resistant. In a suspected bioterrorism attack, patients are commonly given ciprofloxacin until susceptibility testing has been completed. Most nonmodified (i.e., non-weaponized) strains are susceptible to many antibiotics, including imipenem, clindamycin, and erythromycin.

Q: Why do acid-fast bacteria stain poorly with the Gram stain?

A: They have lipids and waxes in their cell walls that do not readily allow the water soluble Gram staining reagents to enter the cells.

Q: What organism, which causes a food-borne disease, demonstrates a characteristic "tumbling" motility when viewed by dark-field microscopy?

    A: *Listeria monocytogenes*

Q: What foods are commonly associated with listeriosis?

    A: Classically, dairy products (milk, cheese, etc.), especially if not pasteurized, but it can also be associated with consumption of meats or even vegetables if they become contaminated

Q: What is the reservoir of *L. monocytogenes*?

    A: The reservoir consists of animals, including cattle, goats, and so on. There are cases of listeriosis due to direct animal contact. The organism can also be found in soil.

Q: Who (patient characteristics) is at highest risk for *L. monocytogenes* infection?

    A: Immunocompromised individuals (transplant patients, people with diabetes, cancer, etc.), the elderly, and pregnant women, including their unborn babies and newborns. Healthy individuals are usually asymptomatic.

Q: How does listeriosis present in at-risk nonpregnant adults and in children (other than newborns)?

    A: Initially, fever, headache, tiredness, and generalized aches and pains. It can progress to septicemia, meningitis, and meningoencephalitis.

Q: How does listeriosis present in pregnant females?

    A: Usually mild flu-like symptoms with low-grade fever

Q: What effect can listeriosis have on the fetus?

    A: It can lead to spontaneous abortion, premature birth, neonatal sepsis, or neonatal meningitis.

Q: Describe the pathogenesis of *Listeria* infection, including a description of its primary virulence factor.

    A: *Listeria monocytogenes* is a gram-positive facultative intracellular parasite, meaning not only can it live within cells, but it can also be grown on artificial laboratory media. During infection, the bacteria bind to receptor sites on the surface of macrophages or epithelial cells. Cell entry is by phagocytosis. The engulfed bacteria in the phagosome are not killed but instead secrete a virulence factor called listeriolysin O (a hemolysin) that breaks down the membrane of the phagosome, thus allowing the organism to enter the cytoplasm. In the cytoplasm the organism can reproduce, and it becomes surrounded by actin filaments. The actin filaments move to one end of the bacterial cells and appear to push them toward the cytoplasmic membrane of the infected macrophage. The cytoplasmic membrane evaginates with the bacterium at the tip of the "pseudopod" and makes contact with another macrophage. The tip of the pseudopod pushes into the adjacent macrophage and is pinched off. The second cell is now infected. The bacterium secretes listeriolysin O, allowing it to escape into the cytoplasm where the whole process repeats itself. The advantage for the *Listeria* is that it can evade the humoral immune response.

Q: A 47-year-old male transient with poor oral hygiene arrived at a free clinic with a soft tissue infection on his lip. Drainage fluid containing small yellow granules were sent to a local microbiology lab for processing. A Gram stain of the crushed granules revealed gram-positive filamentous rods. What is the most likely diagnosis?

    A: Actinomycosis, most likely due to *Actinomyces israelii*. The yellow "sulfur" granules are small bacterial colonies in the exudate.

Q: With respect to oxygen requirements, how is *A. israelii* classified?

    A: Obligate anaerobe

Q: How is *A. israelii* infection acquired?

    A: It is part of our normal flora (oral cavity, upper and lower intestinal tracts, vagina) and gains entry through small (or large) abrasions or cuts in mucous membranes.

Q: What is the drug of choice for *A. israelii*?

    A: Penicillin G

Q: Almost all capsules of bacteria are composed of polysaccharides. What is one exception?

    A: *Bacillus anthracis* has a polypeptide (polyglutamic) capsule. The antiphagocytic capsule is an important virulence factor.

Q: Describe the structure of the toxin of *Bacillus anthracis* and its mode of action.

    A: The exotoxin consists of three protein components, the PA, the LF, and the EF. All three components are secreted by the bacteria during an infection, but none is toxic by itself. During an infection, the PA binds to a surface protein called ATR (anthrax toxin receptor). ATR is on the surface of many cell types; its normal function is unknown. A furin protease, also on the surface of many cell types, cleaves a small peptide (PA_{20}) from PA to yield PA_{63}. Seven PA_{63} molecules associate to form a heptameric prepore. The heptamer then binds three molecules of LF/EF (any combination, including only LF or EF). This prepore–LF/EF complex enters the cell by endocytosis and is present in the cell within a membrane vesicle. Acidification of the vesicle interior causes a conformational change in the toxin complex, allowing EF and/or LF to pass into the cytoplasm of the cell. The EF is a calmodulin-dependent adenylate cyclase and causes an increase in cyclic AMP, resulting in fluid efflux or edema. The LF is a protease that cleaves an intracellular signaling molecule called MAPKK (mitogen activated protein kinase kinase). This inactivation of MAPKK eventually leads to the death of the cell. If this is still confusing, start at the beginning and draw a diagram as you read the answer.

Q: What is Hansen disease?
  A: Leprosy. Hansen disease is the official name in the United States.

Q: Hansen disease primarily affects what tissues/organs?
  A: It is a granulomatous disease affecting the skin and nerves.

Q: What is the treatment for Hansen disease?
  A: Rifampin, dapsone, and clofazimine are taken once per month. Some patients can be cured in as little as six months. Others, especially those with lepromatous leprosy, need to be treated for as long as two years. Rifampin and dapsone should be taken once a month for six months for tuberculoid leprosy. The major thing you need to know is that rifampin and dapsone are the primary drugs and therapy is long term. Recently, thalidomide was approved for the treatment of the cutaneous manifestations of moderate to severe leprosy-associated erythema nodosum.

Q: How is *Mycobacterium leprae* grown in the laboratory?
  A: Normally, it is not. In research laboratories it can be grown in mice (footpads) or in armadillos.

Q: How is Hansen disease spread?
  A: Direct contact with an infectious human

Q: How is TB spread?
  A: Aerosolized droplets (droplet nuclei) from an infected human

Q: For what bacterium can "cold enrichment" be useful in isolating the organism?
  A: *Listeria monocytogenes*. This organism can grow over a wide temperature range, from refrigerator temperatures to near 50°C. Most bacteria are inhibited from growing at cold temperatures. *Listeria monocytogenes* can grow, albeit slowly, and therefore its relative concentration increases compared with other bacteria. This phenomenon is also important in the epidemiology of the disease. If contaminated dairy products are refrigerated, *L. monocytogenes* can grow, resulting in a higher infective dose that is more likely to cause disease.

Q: Here is the obligatory question about what causes rice-associated food poisoning. What is the answer?
  A: *Bacillus cereus*. Because this organism is a spore former, the spores can sometimes survive short-term heating (as with reheated fried rice). This organism can cause food poisoning after ingestion of other foods, too. There are two syndromes: a rapid-onset emetic form (~6–8 hours) presenting primarily with vomiting and a later onset form (~15–20 hours) presenting primarily with a watery diarrhea.

Q: Describe the classification of the mycobacteria.
  A: You probably will not need to know this because it is so confusing, but just in case, the answer follows. Normally, we think of *M. tuberculosis, M. bovis,* and *M. leprae* as individual species. We then take all the others and call them the atypical mycobacteria. The atypical mycobacteria can then be further divided into the (1) photochromogens (yellowish pigment when exposed to light; *M. kansasii, M. marinum,* others), (2) scotochromogens (pigment produced in dark; *M. scrofulaceum,* others; it is unlikely there will be any questions on this group), (3) nonchromogens (nonpigmented; *M. avium* complex, others), and (4) rapid growers (*M. fortuitum/ chelonae* complex; it is unlikely there will be any questions on this group). Note: Other classification schemes have been proposed.

Q: What is the approximate generation time of *M. tuberculosis*?
  A: You do not really need to know the time (about one day in the laboratory under optimal conditions). It is more important that you know that it is one of the more slowly growing organisms in comparison with other bacteria. This slow growth rate accounts for the prolonged time needed to grow the organism in the laboratory (4–8 weeks) and contributes to the prolonged therapy required for patients.

Q: What is the mode of action of isoniazid on *M. tuberculosis*?
   A: It inhibits the synthesis of mycolic acids that are structural components of the cell wall.

Q: What is the cord factor?
   A: It is a cell wall lipid found in *M. tuberculosis* that is a virulence factor for the organism (e.g., it induces the production of tumor necrosis factor and probably has other activities). The cells of *M. tuberculosis* strains that have the cord factor grow in parallel strands (serpentine cords).

Q: Describe the sensitivities and resistances of *M. tuberculosis* to acids, alkalis, disinfectants, ultraviolet light, and heat.
   A: *M. tuberculosis* is generally resistant to acids, alkalis, and disinfectants, but it is susceptible to ultraviolet light and heat.

Q: What is the composition of PPD?
   A: PPD is composed of several proteins and lipids from the cell wall of *M. tuberculosis*. A delayed type hypersensitivity reaction is responsible for a positive skin test.

Q: In a person with no risk factors, what is considered a positive skin test for TB?
   A: An induration of 15 mm or more

Q: What is a positive reaction in a person with HIV disease?
   A: An induration of 5 mm or more

Q: What is a positive reaction in an intravenous drug user?
   A: An induration of 10 mm or more is positive in intravenous drug abusers, persons born in a country with a high TB rate, persons in low socioeconomic groups, residents of long-term care facilities (prisons, nursing homes, etc.), or persons with other diseases such as diabetes and leukemia.

Q: What kind of infection does *Mycobacterium kansasii* cause?
   A: It is similar to that of *M. tuberculosis*.

Q: How do *M. avium* complex infections present in AIDS patients?
   A: The disease is similar to that of *M. tuberculosis*, but it has a more rapid course and disseminates more easily. *M. avium* complex bacteria consist of several subspecies of *M. avium* and are formerly known as *M. avium-intracellulare*.

Q: Why is the BCG vaccine not routinely administered in the United States?
   A: There is a relatively low incidence of TB in the United States. Use of the vaccine converts noninfected persons to PPD positive, thereby decreasing the value of this test for identification of infected individuals. However, the interferon-gamma release assay, as described earlier in this section, is not affected by previous BCG vaccination.

Q: A 17-year-old girl complains of vaginal irritation. On examination, she has a malodorous whitish discharge. Microscopic examination reveals clue cells. What is your diagnosis and treatment?
   A: Bacterial vaginosis due to *Gardnerella vaginalis* and up to 15 or more other bacteria, including *Mobiluncus*, *Prevotella*, and *Porphyromonas*. It is most likely a mixed infection, with *Gardnerella* a common isolate. Treatment is with metronidazole. Note: Both *Gardnerella* and *Mobiluncus* are gram-positive in structure but usually appear gram variable (there are a lot of cells that look gram-negative).

Q: What are clue cells?
   A: Vaginal epithelial cells covered with *G. vaginalis*

Gram-Negative Bacilli

Q: A 22-year-old woman presented to her physician with fever, chills, flank pain, and nausea. Quantitative urine culture yielded 2.22 × 10⁵ cfu/mL of a single organism. She most likely has an infection due to what?

A: *Escherichia coli*. This is always your best guess since *E. coli* is the number one cause of both community-acquired and nosocomial urinary tract infections (UTIs). If additional information were given about specific characteristics of the organism, then the described organism would be the correct answer.

Q: In the previous question, what is the specific disease?

A: Pyelonephritis (kidney infection). Fever, commonly with chills, and flank pain are the most important clues. If she had cystitis only (bladder infection), she would have dysuria, urgency, and increased frequency, usually with suprapubic tenderness, small volume voiding, and pyuria. Some of the signs and symptoms of pyelonephritis and cystitis/urethritis overlap.

Q: Again referring to the same question, if she was asymptomatic but the quantitative culture results were the same (i.e., 2.22 × 10⁵ cfu/mL), would she have a UTI?

A: Yes. Asymptomatic patients with more than 10⁵ cfu/mL of urine are considered to have a UTI. If they are symptomatic, the colony counts can be less and they are still considered to have a UTI. There is some "wiggle room" here, depending on the circumstances, lab procedures, and so on. Not all asymptomatic patients (e.g., geriatric) need to be treated. Pregnant females with asymptomatic UTI should be treated.

Q: Why should pregnant females with an asymptomatic UTI be treated?

A: A UTI in pregnant females can have serious consequences for the fetus, including an increase rate of preterm birth, preeclampsia, and polyhydramnios (increase in amniotic fluid).

Q: A 4-year-old boy recently returned with his family from a camping trip in Arizona. After his return, he developed a high fever and had an enlarged lymph node near his groin. What is the most probable infection?

A: Bubonic plague (explanation to follow next question)

Q: From the previous question, what is the organism?

A: *Yersinia pestis*. In the United States children are usually at highest risk of becoming infected because they are the ones most likely to play with sick or dead animals (because the animals have plague; some infected animals may be resistant and thus asymptomatic carriers). The fleas from the animal jump onto the child and proceed to bite the new host, resulting in infection. A hallmark of infection is the enlarged lymph node, often, but not exclusively, in the groin area. Note: The transmission of *Y. pestis* to new hosts involves the bites of fleas that are infectious because of bacteria trapped in a biofilm on the proventricular valve in the insect's foregut. This causes a blockage that leads to regurgitation of bacteria into the new host as the flea feeds.

Q: Based on the <u>epidemiology</u>, what kind of plague is most consistent in the above case? Based on <u>symptoms</u>, what kind of plague is most consistent in the above case?

A: Sylvatic (based on epidemiology), bubonic (based on symptoms). Sylvatic (non-urban) plague is found in the western United States. Typically, it involves single cases and is not epidemic. Urban plague is spread by urban rodents, usually rats, and is found in some developing countries. It has a high potential to become epidemic. Bubonic plague is characterized by buboes (enlarged lymph nodes). Pneumonic plague (rare in the United States) develops if systemic spread involves the lungs. These patients can spread the organism by coughing. Patients infected via aerosol droplets develop pneumonic symptoms quickly and the mortality rate is higher than in bubonic plague.

Q: What is the most common cause of community-acquired pneumonia among alcoholic patients?

A: Gram-negative bacilli, especially *Klebsiella*, other gram-negative Enterobacteriaceae, *Legionella*, and anaerobes are more common in this group than in nonalcoholics, but *Streptococcus pneumoniae* is the most common in alcoholics and nonalcoholics. Most questions concerning community-acquired pneumonia in alcoholics will be about *K. pneumoniae* (currant jelly sputum), *Legionella* (over age 55, smokers, possibly attended a convention), and anaerobes (aspiration, usually mixed infection with Enterobacteriaceae).

Q: What is the cause of chancroid?

A: *Haemophilus ducreyi*. It is a sexually transmitted disease that is increasing in frequency in the United States. The genital ulcer is soft and painful compared with the hard painless chancre of syphilis. A painful genital ulcer in combination with tender suppurative inguinal adenopathy suggests chancroid.

Q: Enterotoxigenic *Escherichia coli* causes traveler's diarrhea and infant diarrhea. These strains produce two enterotoxins, the heat stable toxin and the heat labile toxin. What is the mode of action of the heat labile toxin?

A: ADP ribosylation of a regulatory (G) protein causing stimulation of adenylate cyclase, resulting in an increase in cAMP concentration. This results in water and electrolyte secretion from intestinal mucosal cells, and thus a watery diarrhea. This is the same mode of action as the cholera toxin.

Q: From the previous question, what is the mode of action of the heat stable toxin?

A: It stimulates guanylate cyclase in small intestinal mucosal cells, resulting in an increase in intracellular cGMP. This prevents mucosal cells from reabsorbing any water and also contributes to the watery diarrhea. There are two forms of heat stable toxins, STa and STb. It is not important for you to know the difference (it is minor).

Q: A 34-year-old male construction worker recently returned from a temporary job in India. Two weeks after returning he developed a headache and fever and became constipated. He told his doctor that he developed the fever several days ago. The doctor took his vitals and noted that, although slightly tachycardic, he had a slower heart rate than expected for his temperature. A blood specimen would most likely be positive for what organism?

A: *Salmonella typhi*, the cause of typhoid fever. Clues include foreign travel to an endemic area, fever and constipation early (can develop diarrhea late), and relative bradycardia (pulse less than that expected for a given temperature). Note: Relative bradycardia is also seen in Legionnaires' disease, pneumonia due to *Chlamydophila* (formerly *Chlamydia*) *pneumoniae*, psittacosis, Q fever, typhus, malaria, babesiosis, leptospirosis, yellow fever, dengue fever, and Rocky Mountain spotted fever. You typically do not need to know any of these except typhoid fever. There are also some noninfectious causes.

Q: Infection by what organisms can result in the hemolytic uremic syndrome?

A: *E. coli*, especially (but not exclusively) strain O157:H7. This is the usual answer, but *Shigella*, which produces the same toxin, and several other species can also cause infections resulting in hemolytic uremic syndrome as a complication.

Q: From the previous question, what is the disease-associated toxin called in *E. coli*?

A: Shiga-like toxin (SLT), or verotoxin. There are actually two Shiga-like toxins, SLT-I and SLT-II, which are nearly identical in structure and have similar modes of action. They are essentially identical to the Shiga toxins of *Shigella dysenteriae*. The term *verotoxin* is an older name referring to the fact that the toxin is cytopathic to tissue culture cells called Vero cells.

Q: What is the mode of action of Shiga and the Shiga-like toxins?

A: These are AB toxins (one A subunit, five B subunits). The A, or active, subunit inhibits protein synthesis by inactivating the 60S ribosomal subunit in susceptible cells. The ensuing cell death results in sloughing of the dead mucosal cells, leading to a bloody diarrhea.

Q: An 18-year-old female college student presented to student health with symptoms consistent with a UTI. Urine culture yielded a *Proteus* species (*P. mirabilis*). What is characteristic about *Proteus* species on agar plates such as blood agar?

A: *Proteus* does not form discrete colonies on many types of media (including blood agar plates). Instead it "swarms" across the plate (it is very motile).

Q: What is yersiniosis?

A: It is an enterocolitis, presenting as diarrhea, fever, and abdominal pain.

Q: What causes yersiniosis?

A: It is caused by *Yersinia enterocolitica* and *Yersinia pseudotuberculosis*, which are found naturally in a number of wild animals. Besides *Y. pestis* (plague) these are the two other important human pathogens within the genus. *Y.*

enterocolitica usually causes diarrhea in children younger than 1 year of age. *Y. pseudotuberculosis* usually does not have a diarrheal component but may present with symptoms of appendicitis. It causes a tuberculosis-like disease in animals.

Q: A 2-year-old girl had rapid onset of fever, headache, stiff neck, and vomiting. Although several organisms can cause such symptoms, what bacterium normally affects only children in this age group (1 month to 5 years)?

A: *Haemophilus influenzae*, especially serotype b (Hib). Other bacteria, such as *Streptococcus pneumoniae*, are becoming more frequent as causes of bacterial meningitis in this age group (because of the success of the Hib vaccine), but if there is an infection due to *H. influenzae* type b, it is most likely to be in this age group, especially if the patient has not been vaccinated with the Hib vaccine. In adults, nontypeable (nonencapsulated) *H. influenzae* is more common in disease (but it is rare). Immunosuppressed adults have a higher risk of Hib infection than immunocompetent adults.

Q: A 37-year-old burn wound patient developed a nosocomial infection of the damaged tissue. She subsequently developed septicemia. Blood culture yielded gram-negative aerobic bacilli that were oxidase positive. A blue-green pigment was observed on brain-heart infusion agar plates. What is the most probable pathogen?

A: *Pseudomonas aeruginosa*. You should *never* miss a question about *P. aeruginosa*. The clues are classic: burn wound, nosocomial infection, gram-negative aerobic bacillus, oxidase positive, blue-green pigment.

Q: Which bacterium is the cause of peptic ulcers and a cause of stomach (gastric) cancer?

A: *Helicobacter pylori*

Q: What causes Legionnaires' disease?

A: *Legionella pneumophila*

Q: Describe the Gram stain morphology of *Legionella pneumophila*.

A: This is kind of a trick question. Although *L. pneumophila* is structurally a gram-negative rod, it stains poorly with the Gram stain. Some modifications in the procedure are useful, such as extending the time for the safranin to 10 minutes. It is better visualized with other stains, such as a silver impregnation or direct fluorescent antibody. On artificial media, it is frequently pleomorphic.

Q: What kind of medium is used to grow *Legionella pneumophila* in the laboratory?

A: Buffered charcoal-yeast extract agar supplemented with L-cysteine and iron

Q: How is legionellosis spread?

A: Aerosolized water droplets from environmental sources such as air conditioning water towers, vegetable misting systems in grocery stores, contaminated indoor water supplies (showers, etc.), and humidifiers. It can live within freshwater amebas in ponds and lakes.

Q: An adventurous pharmacy student and her husband went backpacking over spring break in Brazil. After four days into their trip, they both developed a severe nonbloody diarrhea with "rice-water" stools. What is the most probable organism?

A: *Vibrio cholerae*. Note: In recent years there have been few reported cases of cholera in North or South America. Most reported cases are in Africa.

Q: What is the mode of action of the cholera toxin?

A: It has the same mode of action as the heat labile toxin of *E. coli*. It causes an ADP ribosylation of a G protein in intestinal mucosal cells, resulting in stimulation of adenylate cyclase. The result is the excretion of water and electrolytes. The toxin causes net hypersecretion of intestinal fluid of normal or near-normal chemical composition (protein-free plasma ultrafiltrates containing chiefly water and electrolytes). Microvilli remain intact in cholera. *Vibrio cholerae* is noninvasive, but it adheres to the brush border enterocytes and crypt cells where it secretes the enterotoxin. Flecks of mucus in the stools resemble rice grains. The stools sometimes have a "greenish" coloration.

Q: What causes whooping cough?

A: *Bordetella pertussis*

Q: What are the major virulence factors of *B. pertussis*?

A: The major toxin (the one you should know the most about) is the pertussis toxin, which is an AB toxin (two types of subunits: A = active subunit, B = binding subunit). After one or more of the five binding subunits attach to a ciliated respiratory cell, the active subunit causes ADP ribosylation of a G protein. This results in increased cAMP concentrations, leading to hypersecretion of the respiratory cells. The other toxins include an adenylate cyclase (inhibits phagocytosis plus several other effects) and a tracheal cytotoxin (a peptidoglycan precursor that is toxic to ciliated respiratory epithelial cells). As with most pathogenic gram-negative bacteria, you can assume *B. pertussis* possesses endotoxin and some kind of adhesin. Several other virulence factors have not been mentioned, but you probably do not need to know about them.

Q: What gram-negative bacillus produces red colonies?

A: *Serratia marcescens*. This is an opportunistic pathogen commonly found in nosocomial urinary tract or other opportunistic infections.

Q: A hospitalized 22-year-old woman with a urinary catheter develops a UTI. Gram-negative bacilli were isolated. You suspect either *Pseudomonas aeruginosa* or a member of the family Enterobacteriaceae (*E. coli* and related bacteria). What simple laboratory test allows you to tell these two apart?

A: Oxidase test. *P. aeruginosa* is oxidase positive, and all bacteria in the family Enterobacteriaceae are oxidase negative. In addition, *P. aeruginosa* oxidizes glucose (it is an aerobe and can only grow by respiration, not fermentation), whereas the family Enterobacteriaceae ferments glucose (they are facultative anaerobes and can grow by respiration or fermentation). The oxidase test, not related to the ability to oxidize glucose, is a simple colorimetric test available for detecting cytochrome oxidase (found in oxidase-positive bacteria). (Note: As you go through this series of questions and answers, write out an identification scheme. It will be well worth your time.)

Q: Is *P. aeruginosa* the only oxidase-positive organism?

A: Nope! Actually, most gram-negative bacteria that are not in the family Enterobacteriaceae are oxidase positive. This includes *Campylobacter, Vibrio, Neisseria, Eikenella, Pasteurella, Legionella*, and so on. Except for the Enterobacteriaceae, you do not need to know any other oxidase-negative bacteria.

Q: Now that you know how to differentiate between the Enterobacteriaceae and most other gram-negative rods using the oxidase test, you need to know how to tell some of the bacteria in the family Enterobacteriaceae apart. How can you tell *E. coli* from most other bacteria in the family Enterobacteriaceae?

A: Lactose fermentation. There are other lactose fermenters in the family, such as *Klebsiella*, but you usually do not need to know what they are. They can often be differentiated by other means.

Q: What media can you use to tell lactose fermenters from nonfermenters?

A: Two of the most common are MacConkey agar and eosin methylene blue, or EMB, agar. MacConkey is more common in clinical labs, and EMB is more common in environmental quality (food, water) labs.

Q: What are the three major gram-negative bacteria that cause food poisoning in the United States?

A: *Salmonella, Shigella,* and *Campylobacter*

Q: What media should be plated for the isolation and identification of these three bacteria?

A: MacConkey (usually) or EMB for *Salmonella* and *Shigella* and a selective medium (you do not need to know what it is) for *Campylobacter*. You should know, however, the incubation conditions for growing *Campylobacter* (see the next question). For most specimens sent to the clinical microbiology lab for culture, a nonselective blood agar plate is usually inoculated.

Q: What are the incubation conditions for *Campylobacter*?

A: The important condition for optimal growth is decreased oxygen concentration, because *Campylobacter* is microaerophilic (you do not need to memorize the percent of oxygen, but it is around 5%). They also grow better at 42°C than at 37°C.

Q: What is the major species of *Campylobacter* and what major disease does it cause?

A: *Campylobacter jejuni* is a major cause of gastroenteritis. Evidence suggests that it is the most common cause of bacterial gastroenteritis worldwide.

Q: What is the major source of *C. jejuni* infections?

A: Most are due to contaminated poultry.

Q: What does *Campylobacter* look like microscopically?

A: Small gram-negative curved rods

Q: *Campylobacter* gastroenteritis is a predisposing factor for what important neuropathy?

A: Guillain-Barré syndrome (and its less common variant, Miller Fisher syndrome). This is an example of molecular mimicry. The immune system makes antibodies to a component of the lipo-oligosaccharide of the *Campylobacter* cell wall. These antibodies cross-react with gangliosides in peripheral nerves, resulting in demyelination.

Q: What other genera of bacteria are curved rods?

A: *Vibrio* and *Helicobacter*. Curved rods, or vibrios, are sometimes described as spiral shaped. This is especially true of *Campylobacter* and *Helicobacter* because sometimes as they grow and divide, the daughter cells do not break apart. If this continues, the arrangement resembles spiral-shaped bacteria, but they are just short chains of curved rods stuck together.

Q: Describe the colonies that grow on MacConkey agar if *Salmonella* or *Shigella* is present in a stool specimen.

A: The colonies are colorless, that is, they are not red or pink. Colorless colonies indicate lactose nonfermenters (lactose negative). Red or pink colonies are lactose fermenters (lactose positive), primarily *E. coli*. *E. coli* is normal flora, so it is always there. Therefore in a stool specimen with *Salmonella* or *Shigella* you should see colorless and red (or pink) colonies (the *E. coli*). The colorless colonies could be other lactose nonfermenters besides *Salmonella* or *Shigella*; therefore the bacteria must be identified.

Q: How are *Salmonella* and *Shigella* differentiated from each other?

A: *Salmonella* is hydrogen sulfide positive; *Shigella* is hydrogen sulfide negative. This is easy to tell by growing the bacteria in media (like triple sugar iron) that contain iron salts. Hydrogen sulfide turns the medium black. In addition, *Shigella* is nonmotile.

Q: What other bacteria are nonmotile?

A: This seems like a tough question at first, but the answer is surprisingly easy. Just assume all bacilli are motile *except Shigella, Yersinia,* and *Klebsiella*. Look at the first letter of each. These bacteria are too SYK (pronounced "sick") to move. In addition, all cocci, regardless of their Gram reaction, can be considered as nonmotile. There are exceptions to these rules, but you do not need to know them.

Q: The only other thing you might need to know is how to tell *E. coli* from the other lactose fermenters. How?

A: *E. coli* is indole positive; many others (there are exceptions you do not need to know about) are indole negative. How can you remember this? *E. coli* ends with an "i" and indole starts with an "i."

Q: *E. coli* can also cause gastroenteritis. How is it isolated and identified?

A: Most cases of *E. coli* gastroenteritis are not identified by culture because it is usually a mild self-limited disease such as traveler's diarrhea. These mild infections are commonly due to so-called enterotoxic *E. coli*. Occasionally, such as severe gastroenteritis due to *E. coli* O157:H7, culture and identification with special media is required.

Q: What is another name for *E. coli* O157:H7?

A: Enterohemorrhagic *E. coli*

Q: What disease does *E. coli* O157:H7 cause?

A: It can cause hemorrhagic colitis that can lead to hemolytic uremic syndrome (HUS) as a complication. Patients with HUS present with microangiopathic hemolytic anemia, thrombocytopenia, and renal dysfunction. The onset is rapid. It occurs in children about one week after the onset of gastroenteritis. HUS may occur in adults, most commonly complicating pregnancy or the postpartum period. Acute renal failure develops in 60% of children with HUS and usually resolves in weeks with supportive therapy only. The role of antibiotics is controversial.

Q: *Vibrio parahaemolyticus* causes what kind of disease?

A: Although it occasionally is a cause of seawater-associated sepsis, it is more commonly a cause of seafood-associated food poisoning (e.g., oysters, crabs, shrimp, etc.). It is one of the most common causes of gastroenteritis in Japan but is usually mild.

Q: In *Haemophilus influenzae* type b, "type b" refers to what?

A: Serotype b of the capsule. There are other serotypes, but type b is most common in human disease.

Q: Which age group is usually afflicted by pertussis?

A: Many physicians consider pertussis a childhood disease, affecting primarily infants younger than age 2 years. However, it is estimated that 20–30% of coughs lasting more than two weeks in adults is due to *Bordetella pertussis*. This is the primary reason that the Tdap vaccine has been recommended for adult boosters every 10 years.

Q: A UTI due to urease-positive bacteria such as *Proteus* or *Klebsiella* can result in what kind of complication?

A: Stone formation. An alkaline urinary pH (due to NH_4^- accumulation from the splitting of urea) is conducive to the formation of struvite stones.

Q: *E. coli* (along with *Streptococcus agalactiae*) is considered one of the most common causes of neonatal meningitis. How do infants become infected?

A: As they pass through the birth canal. Both of these organisms are part of the normal vaginal microflora.

Q: What role does the lipopolysaccharide (LPS) of gram-negative bacteria play in virulence?

A: It is the endotoxin of these bacteria and is released as the bacteria die and lyse. It contributes to fever, endotoxic shock, disseminated intravascular coagulation, and so on. Many of the effects of LPS are cytokine mediated, with interleukin-1 and tumor necrosis factor being particularly important.

Q: What is the number one cause of respiratory infections in cystic fibrosis patients?

A: *Pseudomonas aeruginosa*, especially in older patients. The strains isolated from cystic fibrosis patients typically are very mucoid in culture because of their large capsule. The capsule facilitates adherence to mucous membranes.

Q: Curved motile rods that are seen in a wet mount of a stool specimen from a patient with severe watery diarrhea are suggestive of what?

A: Cholera. Culture is preferable for diagnosis.

Q: What is the most common species of *Shigella* in the United States?

A: *Shigella sonnei*

Q: What is the reservoir for *Shigella*?

A: All species are found in humans only. *Shigella* should not be considered as normal flora. If it is present, the host is either sick or an asymptomatic carrier. *Salmonella* is found in animals.

Q: In general, how does the infective dose (the number of bacterial cells required to cause an infection) compare between *Salmonella* and *Shigella*?

A: It takes far fewer cells of *Shigella* (~50–200 cells) to cause gastroenteritis compared with *Salmonella* and most other bacteria (>100,000 cells). The reason for this is that *Shigella* is more resistant to the acidity of the stomach than are most other bacteria.

Q: How does *Shigella* gastroenteritis present?

A: One to three days after ingestion, the patient develops fever, nausea, and a watery diarrhea. It can then develop into a bloody diarrhea. All four species of the genus *Shigella* (*S. dysenteriae*, *S. flexneri*, *S. boydii*, and *S. sonnei*) cause disease by secretion of an enterotoxin (Shiga toxin) after invasion of intestinal epithelial cells. *Shigella* typically invades epithelial cells of the small intestine, then they invade M cells (microfold cells) of the large intestine. They multiply intracytoplasmically, killing cells and causing small ulcers, resulting in marked inflammation. This process is usually accompanied by mucosal bleeding and excess mucous secretion. The inflamed bowel causes pain and cramps. Most cases of shigellosis affect children. The lamina propria is often involved in infection. The Shiga toxin is involved in the disease by inhibiting protein synthesis, leading to cell death.

Q: What color are *Shigella* colonies on MacConkey agar?

A: Colorless. They are nonlactose fermenters.

Q: What is the treatment for shigellosis?

A: Usually just fluid and electrolyte replacement. In severe cases, antimicrobials such as a fluoroquinolone (preferred) or trimethoprim-sulfamethoxazole might be used.

Q: How many species of *Salmonella* cause gastroenteritis?

A: Only one species (*S. enterica*), but there are more than 2500 different serovars, including *S. typhimurium* (officially [CDC and WHO] designated as *S. enterica* subspecies *enterica* serovar Typhimurium) and *S. enteritidis* (officially designated as *S. enterica* subsp. *enterica* serovar Enteritidis). Most serotypes are not named but have serotype designations. For convenience, most microbiologists use the classical taxonomic names for named species, i.e., *S. typhimurium*, *S. enteritidis*, *S. arizonae*, etc.

Q: What bacteria cause enteric fever?

A: The most severe form is due to *Salmonella typhi* (officially designated as *S. enterica* subsp. *enterica* serovar Typhi). The disease is typhoid fever. *Salmonella paratyphi* and *Yersinia enterocolitica* cause milder forms of enteric fever. *Y. enterocolitica* is also a common cause of pediatric diarrhea.

Q: What is the reservoir for *S. typhi* and how is infection acquired?

A: It is a strict human pathogen spread by the fecal–oral route.

Q: Describe the pathogenesis of *S. typhi*.

A: *S. typhi* is a facultative intracellular parasite that infects macrophages in the Peyer patches. The organism then spreads via the blood to other organs such as the liver, gallbladder, spleen, and bone marrow. There is then a second bacteremic spread (1–2 weeks after ingestion) that results in reinfection of the intestinal tract. The primary difference between *S. typhi* and the *Salmonella* serovars that cause gastroenteritis is that *S. typhi* has a significant bacteremic stage. Bacteremia is uncommon in gastroenteritis unless the patient is immunocompromised or has other predisposing factors.

Q: How does typhoid fever typically present?

A: Within about 1–2 weeks after infection, the patient experiences fever (40°C), followed by chills, malaise, anorexia, myalgias, arthralgia, cough, sore throat, and headache. There may be some mental confusion. Relative bradycardia (not as tachycardic as expected for the degree of fever) is common. Rose spots may appear (look at some images), especially on the abdomen and chest, and are due to the multiplication of the organism in the skin. Note: Other infections with relative bradycardia include Legionnaires' disease, Rocky Mountain spotted fever, leptospirosis, psittacosis, Q fever, typhus, malaria, babesiosis, yellow fever, and dengue fever. In most cases for standardized exams, you do not need to know these.

Q: What is the therapy for typhoid fever?

A: A fluoroquinolone or a third-generation cephalosporin such as ceftriaxone is the drug of choice.

Q: Is there a vaccine available for typhoid fever?

A: There are two vaccines recommended by the CDC for travel to areas where *S. typhi* is present: an oral live cell attenuated and a conjugated vaccine based on the Vi antigen (intramuscular).

Q: What is the Vi antigen of *Salmonella*?

A: It is the polysaccharide capsule and is a virulence factor for *S. typhi*. Vi-positive strains are more resistant to complement-mediated killing and phagocytosis than are Vi-negative mutants. It is absent in most *Salmonella* serovars but is present in all isolates of *S. typhi* from patients with acute infections and all asymptomatic carriers. The LPS of *S. typhi* is also an important virulence factor.

Q: From what infectious disease did Anne Frank die?

A: Typhus fever, most likely epidemic louse-borne typhus due to *Rickettsia prowazekii* (you don't really need to know the answer to this question; see note below). The infection is associated with crowded unsanitary living conditions such as in concentration or some refugee camps. The disease today is most prevalent in mountainous areas of Africa, Asia, and Latin America. Anne Frank died in March 1945 at the Bergen-Belsen concentration camp at the age of 15, just two weeks before the camp was liberated.
Note: It is unusual to have a question over this disease on most nationally standardized exams. Typhus fever will most likely be listed as one of the wrong answers, such as a question about typhoid fever.

Q: Describe the pathogenicity of *Campylobacter jejuni*.

A: *C. jejuni* has several virulence factors that appear to contribute to the diarrhea. It produces a heat labile enterotoxin similar to that of enterotoxigenic *E. coli* that causes an increase in cAMP. This can result in a watery diarrhea. It also invades the lamina propria in the small and large intestines, which may contribute to the bloody diarrhea. A cytotoxin in many strains may also be contributory. The polar flagella and other surface components appear to contribute to intestinal adherence. On a worldwide basis, most infectious diarrheas are due to *C. jejuni*, enterotoxigenic *E. coli,* and rotavirus.

Q: Describe the pathogenesis of *Salmonella* food poisoning.

A: After ingestion, the bacteria adhere to the microvilli of the small intestine where they cause ruffling. The bacteria invade both intestinal epithelial cells and macrophages. At least two enterotoxins have been described in *Salmonella*. One of the enterotoxins causes the elevation of cAMP in cells after the *Salmonella* invade the intestinal epithelial cells, resulting in hypersecretion (watery diarrhea). The toxin is injected into the intestinal epithelial cells by a type III secretion system. The role of the other enterotoxin is not clear. The infection is usually localized with no bacteremia. In rare cases they can cause systemic infections, especially in immunocompromised patients (AIDS, etc.). It is important to remember, however, that most food poisoning organisms can cause bacteremia and systemic involvement, such as osteomyelitis in immunocompromised people, although it is uncommon.

Q: How does *Salmonella* get inside of eggs?

A: They infect the ovaries of hens and the shell is added after the egg is infected.

Q: What do colonies of *Klebsiella* look like?

A: They are very mucoid compared with most other bacteria.

Q: Where in the United States is plague most common?

A: In what is called the "four corners" area, where Utah, Colorado, New Mexico, and Arizona meet at the same spot. Of these, it is most common in New Mexico and Arizona. The disease can be found in other states in the region, especially California, Oregon, and Nevada.

Q: Historically, why was the plague referred to as the Black Death?

A: Systemic spread results in a darkening of the skin because of subcutaneous hemorrhages.

Q: What is the reservoir for *Vibrio cholerae*?

A: *V. cholerae* is a strict human pathogen. Infection occurs via the fecal–oral route.

Q: What is the typical therapy for uncomplicated cholera?

A: Fluid and electrolyte replacement (this is the most important). The use of antibiotics such as a tetracycline (the drug of choice), a fluoroquinolone, or trimethoprim-sulfamethoxazole can help speed recovery and limit the spread to others.

Q: A businessman from Omaha was snorkeling while on vacation in Cozumel, Mexico. A scrape he received from some coral later appeared to be infected and showed evidence of necrosis. He was placed on a tetracycline, and although he was very ill, he survived the infection. What is the most likely pathogen?

A: *Vibrio vulnificus*. Think of this organism in sepsis associated with seawater. Wound infection with *V. vulnificus* can have a mortality rate of more than 50%, especially if treatment is delayed or the patient is immunocompromised. *V. vulnificus* is quite virulent, with some estimates that as few as 100 cells can cause a wound infection. It occasionally can cause a gastrointestinal infection that has a much lower mortality rate. GI infections can be acquired after ingestion of raw oysters or contaminated clams, crabs, and other shellfish. A related organism, *V. parahaemolyticus*, is more common as a gastrointestinal pathogen but conversely can occasionally cause life-threatening wound infections. Both species are halophiles.
Note: Of 17 *Vibrio*-associated wound infections associated with Hurricane Katrina in 2005, 14 were due to *V. vulnificus* (three deaths) and 3 were due to *V. parahaemolyticus* (two deaths).

Q: *Haemophilus influenzae* type b was eventually isolated from the cerebrospinal fluid of a 2-year-old nonvaccinated boy. The organism did not grow on typical laboratory media. Why not?

A: It requires two nutritional factors for growth, the X factor (heme) and the Y factor (NAD). Both need to be added to artificial media (usually chocolate agar) as a supplement.

Q: What is the reservoir for *H. influenzae* type b?

A: This is another organism that is a strict human pathogen. It is spread by droplets, fomites, or direct contact.

Q: What are the virulence factors for *H. influenzae* type b (Hib)?

A: The antiphagocytic polysaccharide capsule (composed of polyribitol phosphate) and the secretion of IgA protease are the two most prominent. For this and all other gram-negative bacteria, you can assume the LPS has a role in virulence. Most gram-negative bacteria, including *H. influenzae*, have pili that function in adherence.

Q: What age group is at highest risk for Hib infection?

A: Generally, young children from about 3 months to 3 years of age, especially if they have not been vaccinated. Because of the use of the Hib vaccine, infections have been decreasing. Many cases of bacterial meningitis in this age group are due to other bacteria, such as *Streptococcus pneumoniae*. These other infections are not necessarily increasing; they just appear more common relative to Hib infections. Hib can also cause infections in older adults, especially those with a waning immune system.

Q: How do patients with Hib meningitis present?

A: Patients (except infants) with bacterial meningitis, regardless of the organism, usually have similar symptoms such as stiff neck, headache, fever, irritability, and vomiting. Cerebrospinal fluid must be sent to the laboratory for specific identification of the organism.

Q: What is the prognosis of a child with Hib meningitis?

A: There is a mortality rate of about 5%. Among those who survive, there is a significant risk of neurological sequelae such as hearing loss, convulsions, and mental retardation.

Q: How is Hib infection prevented?

A: The Hib conjugate vaccine should be given at 2, 4, and 6 months and again at 12–15 months. There may be some variation in this schedule depending on which vaccine formulation is given. The DTaP–Hib combination products should not be used for primary vaccination in infants at ages 2, 4, or 6 months because there may be a lower immune response to the Hib vaccine component. In recent years there have been fewer than 20 cases a year in the United States.

Q: What other infections involve strains of *H. influenzae*?

A: *H. influenzae* was a common cause of septic arthritis, epiglottitis, and cellulitis in children, but since the advent of the Hib vaccine, these infections have also been on the decrease. Nonencapsulated strains that are not affected by the vaccine (the vaccine is capsular based) can cause infections such as conjunctivitis, otitis media, and sinusitis. The two most common bacterial causes of otitis media and acute rhinosinusitis are *H. influenzae* and *Streptococcus pneumoniae*. Viruses are the most common cause of acute rhinosinusitis.

Q: How is infection with *Pasteurella multocida* acquired?

A: Primarily from cat or dog bites and scratches. This is probably the only thing you need to know about this organism.

Q: Why should you know about *Aeromonas*?

A: Although not typically discussed to any extent in most pathogenic microbiology courses, *Aeromonas*, especially *A. hydrophila*, is an emerging pathogen causing an array of diseases, including gastroenteritis; hemolytic uremic syndrome; urine, wound, and burn infections; necrotizing fasciitis; pneumonia; meningitis; and eye infections. *Aeromonas* is a gram-negative, oxidase-positive organism. Species can be found in many environments but are best known for being found in water, even if chlorinated. There is at least one case of epididymitis and pyelonephritis associated with intercourse in a pool.

Q: A hunter developed a necrotic lesion on a knuckle six days after skinning a rabbit. What is the most probable infecting agent?

A: *Francisella tularensis*. This is the most common scenario for the development of tularemia. This infection is usually your best guess when a rabbit is the source of the infection. Rabbits can also be carriers of *Yersinia pestis* (plague), but the infection presents differently.

Q: Are there any other routes of infection by *F. tularensis*?

A: Yes, almost any route you can think of: direct contact (as with skinning a rabbit), tick bite, deer fly bite, inhalation, and ingestion. Because of the many different ways of becoming infected and the very low infectious dose (as few as 10 cells by direct contact), *F. tularensis* is considered a possible bioterrorist weapon. In suspected cases of tularemia, many laboratories send the specimens to the CDC for identification rather than risk infection of laboratory personnel. A vaccine is available for persons at risk. *F. tularensis* is a CDC category A agent.

Q: What are some virulence factors of *Pseudomonas aeruginosa*?

A: Two major ones are the antiphagocytic polysaccharide capsule and toxin A. Others include a type III secretion system and various enzymes such as proteases, lipase, and lecithinase that are spreading factors.

Q: What is the mode of action of toxin A of *P. aeruginosa*?

A: It has the identical mode of action as the diphtheria toxin. Try to recall what that is before reading further. The mode of action is that both of them cause an ADP ribosylation of elongation factor 2, resulting in an inhibition of protein synthesis. Although both toxins have the same mode of action, they are not related to each other.

Q: *P. aeruginosa* can produce a distinctive odor. How is the odor described?

A: Sweet or fruity are the two most common descriptions, but others such as grape-like or corn tortilla–like have been used. The description of the odor is highly subjective.

Q: List some infections that can involve *P. aeruginosa*.

A: The following is a partial list of the opportunistic infections that can involve *P. aeruginosa*: catheter-associated and nonassociated UTIs, pneumonia, burn wounds, endocarditis (especially in intravenous drug users), immunosuppressive drug associated, dermatitis (hot tubs, whirlpools), eye infections (contact lens solutions, postsurgical), swimmers ear, bacteremia, and respiratory (especially cystic fibrosis patients). Many of these infections can be due to other bacteria such as *E. coli*, *Staphylococcus aureus*, *S. epidermidis*, and *Proteus*.

Q: What is the drug of choice for *Pseudomonas aeruginosa* infections?

A: For *P. aeruginosa* infections, susceptibility testing is necessary because of the high frequency of drug resistance. Therefore this question is not likely to be on national examinations. If this question is asked, guess ciprofloxacin for a UTI. Other infections require combination therapy (e.g., piperacillin-tazobactam or ticarcillin-clavulanate) and are even less likely to be asked.

Q: What infections involve *Stenotrophomonas maltophilia*?

A: *Stenotrophomonas* (formerly *Xanthomonas*) *maltophilia* is a multidrug-resistant nosocomial pathogen causing infections similar to *Pseudomonas aeruginosa*. It is an opportunistic pathogen, infecting patients who are severely debilitated or immunosuppressed. It is frequent in patients with conditions such as cystic fibrosis, malignancy, or mechanical ventilation. Infections can be difficult to treat because the organism is usually multidrug resistant. Other organisms, such as *Burkholderia* (*Pseudomonas*) *cepacia* can also cause these kinds of infections.

Q: A 67-year-old man presented with severe bacterial pneumonia including tachypnea, rales, nonproductive cough, fever, diarrhea, vomiting, and anorexia. He has smoked cigarettes since the age of 16. The bacterium did not grow on typical laboratory media but did grow on a medium containing cysteine and iron. How would the organism appear on Gram stain?

A: It would be difficult to see. The organism is *Legionella pneumophila* (or other related *Legionella* species) and stains poorly with the Gram stain. Structurally, it has a gram-negative architecture.

Q: How is *L. pneumophila* stained for microscopic observation?

A: Special methods such as fluorescence microscopy or a silver impregnation stain reveals the thin pleomorphic bacilli.

Q: The 67-year-old man above visited a close friend in the hospital who had been diagnosed with legionellosis (Legionnaires' disease). He suspected that he caught this infection from his friend. As his health care provider, what do you tell him about how this infection is acquired?

A: You tell him that this infection is not spread from person-to-person. The usual sources are associated with water droplets: air conditioning units, contaminated plumbing, vegetable sprayers in grocery stores, and so forth.

Q: What methods are used to remove *Legionella* from plumbing systems?

A: Superheat ($\geq 80°C$) and hyperchlorinate the water

Q: How is pertussis spread?

A: *Bordetella pertussis* is a strictly human pathogen spread primarily by droplets.

Q: List some virulence factors of *B. pertussis*.

A: You will not be expected to know them all (more than 10 have been identified). The primary ones you should recognize are toxins and include the pertussis toxin, adenylate cyclase, and tracheal cytotoxin. Others include type III secretion and various adhesins.

Q: From the previous question, what are the modes of action of each toxin that is listed?

A: Pertussis toxin: ADP ribosylation of the G_1 protein of host cells resulting in an elevation of cAMP. Adenylate cyclase: inhibition of phagocytosis (killing). Tracheal cytotoxin: kills ciliated respiratory epithelial cells (it is a peptidoglycan precursor)

Q: Describe the symptomatic stages of pertussis.

A: Pertussis goes through two main stages, the initial catarrhal stage characterized by "cold-like" symptoms (coughing, sneezing, fever, malaise, etc.) and the following paroxysmal stage (about 10 days later) in which the characteristic whooping cough is prominent.

Q: How is pertussis prevented?

A: Vaccination with the DTaP vaccine (2 months to 6 years) or Tdap (adolescents and adults). The "aP" and "ap" refers to the acellular *Bordetella pertussis* components (the composition varies but consists of components such as pertussis toxoid and filamentous hemagglutinin).

Q: A 42-year-old male butcher in Montana would occasionally cut and wrap meat from elk and other wild animals that local hunters had killed. Recently, he experienced a rapid unplanned weight loss (12%). He also had generalized symptoms such as myalgia, severe headache, intermittent fever, and increased sweating. Physical examination revealed enlarged spleen and lymph nodes. His wife reported that he experienced recent episodes of confusion and depression. What bacterial infection would you expect?

A: Brucellosis. This infection is most common in developing countries. Pasteurization of dairy products and vaccination of livestock have contributed most to the low incidence in the United States (100–125 reported cases a year). The bacteria are found in the United States in bison (primarily) and elk (*Brucella abortus* in both) in and around Yellowstone and Grand Teton National Parks and in feral swine (*Brucella suis*) in the southeast U.S. Occasionally, cattle herds become infected. Persons who have contact with these animals are at risk. In developed countries it is most often seen in foreign travelers returning from endemic areas (usually associated with consumption of unpasteurized milk or soft cheeses). Intermittent (undulant) fever is seen in many (but not all) patients. In humans, the disease is also called undulant fever or Malta fever.

Q: **What is the cause of cat scratch disease?**
A: *Bartonella henselae*. This infection is actually quite common, with an estimated 20,000–25,000 cases in the United States each year. Most infections are in children and young adults. It presents primarily with enlarged regional lymph nodes (may be tender) and fever. A related species, *Bartonella quintana,* is a cause of bacillary angiomatosis in AIDS and immunosuppressed patients and in trench fever (common in World War I).

Q: **What are the incubation conditions for *Helicobacter pylori* in the laboratory?**
A: You do not need to know the growth medium, but you should know that it is microaerophilic, as is *Campylobacter*. Therefore it is grown in about 5% oxygen (do not memorize the 5%; just know it is lower that atmospheric oxygen).

Q: **What are the virulence factors for *H. pylori*?**
A: An important virulence factor is the secretion of urease, because the urease allows the organism to survive and multiply in the low acidity of the stomach. The enzyme breaks down urea, producing ammonia that when dissolved in water forms the ammonium ion, resulting in an increase in the local pH in the stomach. If it was not urease positive, it would not survive long in the acidity of the stomach. The organism is also motile (polar flagella), allowing it to swim toward the gastric mucosa where it can attach. *H. pylori* have other important virulence factors, such as outer inflammatory protein A, vacuolating cytotoxin, cytotoxin-associated antigen, and colonization factors.

Q: **How is infection with *H. pylori* diagnosed?**
A: There are a number of different ways, some more pleasant for patients than others:

- The ^{13}C- and ^{14}C-urea breath tests rely on the ability of the *H. pylori* urease to break down ingested labeled urea to the ammonium ion and ^{13}C- or ^{14}C-labeled carbon dioxide that is exhaled and can be detected. The breath test is noninvasive and has a high sensitivity (90–100%) and specificity (89–100%). It is also the most expensive.
- Serology (e.g., enzyme-linked immunosorbent assay, others) can detect bacterial antigens (best for initial screening). A stool antigen assay is also commonly available and is comparable to the breath test for sensitivity and specificity.
- Histology of biopsied tissue is an invasive method (endoscopy) that reveals the bacteria closely adherent to the gastric mucosa. This and the next two methods have a lower sensitivity because only a small area is sampled.
- A urease test can also be performed on the biopsied tissue.
- Culture (from biopsy) is also an option. It can have a lower sensitivity but is highly specific.

Q: **What is the therapy for *H. pylori* infection?**
A: A combination of a proton pump inhibitor + clarithromycin + either metronidazole or amoxicillin. An alternative is bismuth subsalicylate + metronidazole + tetracycline + either a proton pump inhibitor or H$_2$ blocker. Combination therapy is always used.

Q: **A 45-year-old alcoholic developed a pneumonia in which she produced a thick sputum that became bloody later in the infection. The lab isolated a gram-negative bacillus that produced very mucoid colonies on blood agar plates. The organism was nonmotile and urease positive. What is the most likely organism in this disease?**
A: *Klebsiella pneumoniae*, which is an opportunistic pathogen. There are lots of clues in this question: alcoholic (immunosuppression), pneumonia, thick sputum (due to capsule), bloody sputum (necrosis), gram-negative, mucoid colonies (due to capsule), nonmotile (*Shigella* and *Yersinia* are other major pathogens in the family Enterobacteriaceae that are nonmotile), and urease positive (*Proteus* is another common urease positive gram-negative bacillus).

Q: In general terms, what are the basic principles for treating anaerobic infections?

A: Antibiotics and drainage (or debridement). Drainage or debridement of an abscess is not always possible (e.g., septicemia, pseudomembranous colitis, bacterial vaginosis, etc.). Hyperbaric oxygen, if available, can be an important adjunct to treatment.

Q: A blood culture from a patient with a serious anaerobic infection yielded gram-positive "boxcar"–shaped bacilli showing a double zone of hemolysis on blood agar plates. What is the most probable organism?

A: *Clostridium perfringens.* The description given is classical for this organism. The inner zone of hemolysis is complete (beta-hemolysis) and the outer zone is partial (a lighter shade of hazy pink, not green as in alpha-hemolysis).

Q: What organism causes tetanus?

A: *Clostridium tetani*

Q: What is the drug of choice for suspected tetanus?

A: Metronidazole. It prevents further growth of the organism so that no additional toxin is produced. Penicillin G (the classic choice) or doxycycline are a good alternatives.

Q: A medical student was trying to identify a laboratory unknown. He knew it was a gram-negative obligately anaerobic rod. His lab instructor told him that this species was found as normal flora in the colon and was the most common anaerobe found in serious anaerobic infections. What is the most likely identification of this organism?

A: *Bacteroides fragilis*

Q: What is the drug of choice for *B. fragilis*?

A: Metronidazole

Q: Why can't penicillin G be used for *B. fragilis*?

A: It constitutively produces penicillinase.

Q: List three incubation systems that allow for the growth of anaerobic bacteria.

A: Anaerobic glove box (or chamber), jar, and pouch (or biobag)

Q: A hospitalized cancer patient receiving chemotherapy was being treated with antimicrobials for recurrent nosocomial infections. He later developed a prolonged diarrhea. Endoscopic examination demonstrated the presence of a yellowish-white pseudomembrane on the wall of the colon. Withdrawal of antimicrobials had little effect on the diarrhea or the pseudomembrane. What antibiotic should now be used?

A: Oral metronidazole, because the patient probably has antibiotic-associated pseudomembranous colitis due to *Clostridium difficile.* Oral vancomycin is a good alternative.

Q: Anaerobic bacteria can cause infections at virtually any anatomical site. In most cases of infections due to anaerobes, what kinds of specimens must be avoided?

A: Any specimen contaminated with normal flora. Anaerobes are the most common normal flora organisms at virtually all sites that have a normal microbial flora.

Q: Many anaerobic bacteria can survive in the presence of oxygen until anaerobic conditions develop. Name two enzymes that these bacteria may possess that allow for their survival in the presence of oxygen.

A: Superoxide dismutase and catalase are two of the most common enzymes, but there may be others such as various peroxidases in some species.

Q: A 37-year-old woman presented with abdominal pain, weight loss, vaginal discharge, and fever. An intrauterine device she had been using for birth control for several years was removed and sent to the laboratory for culture. The organism that was isolated from the surface of the intrauterine device was a gram-positive obligately anaerobic branching bacillus. What is the most probable organism (genus and species) and what is the drug of choice?

A: *Actinomyces israelii* is the most likely organism, and penicillin G is the drug of choice. The major clue is that the infection is associated with the use of an intrauterine device, a common surface for the growth of this organism.

Q: What are the hallmarks of an anaerobic infection?

A: Foul smell (due to fermentation and the release of volatile fatty acids) and necrosis

Q: What is the mode of action of the tetanus toxin?

A: It is a neurotoxin that ascends (retrograde transport) the motor nerves from the infection site to the anterior horn cells of the spinal cord where it blocks the release of glycine and gamma-aminobutyric acid, which are both neurotransmitters. In affected patients, it causes a spastic paralysis.

Q: What is the mode of action of the botulism toxin?

A: It is a neurotoxin that blocks the release of acetylcholine at cholinergic nerve synapses. Affected patients exhibit a flaccid paralysis.

Q: What is the most common cause of gas gangrene?

A: *Clostridium perfringens*, but about 20 other species of clostridia can cause the infection

Q: What is a primary toxin involved in gas gangrene due to *C. perfringens*?

A: The alpha toxin, which is a lecithinase. This toxin can be detected using egg yolk agar. Egg yolks are high in lecithin. The alpha toxin partially breaks down the lecithin, resulting in an insoluble precipitate in the agar.

Q: What is the genus name of the most common gram-positive anaerobic coccus found in human infections?

A: *Peptostreptococcus*. It is not necessary to learn species names for this organism. Many clinicians refer to these bacteria as anaerobic streptococci.

Q: What organism is most commonly associated with acne (acne vulgaris)?

A: *Propionibacterium acnes*. This organism does not cause acne; it contributes to the inflammatory response that results in the inflamed hair follicles and sebaceous glands.

Q: Where is the location of the gene that codes for the botulism toxin (plasmid, lysogenic bacteriophage, or chromosome)?

A: Lysogenic bacteriophage. This is an example of lysogenic conversion. Strains that are nonlysogenic do not produce the toxin.

Q: Where is the gene that codes for the tetanus toxin (plasmid, lysogenic bacteriophage, or chromosome)?

A: Plasmid

Q: Tetanus spores are ubiquitous, but tetanus is not common in the United States. Why?

A: Most people have been vaccinated with the DPT, DTaP, or Tdap vaccines.

Q: What is the composition of the component of the above-listed vaccines that results in tetanus immunity?

A: Tetanus toxoid

Q: What is a toxoid?

A: It is a bacterial toxin that has been inactivated so that it is no longer toxic. However, it is still immunogenic and can confer immunity to the tetanus toxin. The tetanus and diphtheria toxins are converted into toxoids in the above vaccines by treatment with formaldehyde.

Q: What is trismus?

A: Lockjaw, due to the effect of the tetanus toxin on the masseter muscles. It is a common feature of tetanus.

Q: What is the name given to the condition in tetanus that results in continuous contraction of facial muscles?

A: Risus sardonicus

Q: How is tetanus diagnosed?

A: Diagnosis is based on clinical signs and symptoms.

Q: What is the therapy for tetanus?

A: Cleansing of the wound (although the site of entry of the spores is often not apparent), metronidazole (to inhibit further growth of the bacteria), and tetanus immune globulin (binds to and inactivates any circulating toxin). For patients with a puncture wound and the immune status is unknown, tetanus toxoid should be administered. Penicillin G and doxycycline are alternative antimicrobials.

Q: What is the source of most anaerobic infections?

A: Most are due to normal flora.

Q: Long-term use (actually, as few as two days) of antibiotics can result in what anaerobic infection? What is the organism?

A: Antibiotic-associated diarrhea or colitis due to *Clostridium difficile*. The disease is usually hospital-associated, but community-acquired cases are increasing. A hypervirulent strain of *C. difficile* is emerging in the United States, Europe, and Japan. Although previously not common, there appears to be an increase in cases after the use of fluoroquinolones.

Q: What anaerobes can cause illness and are not necessarily due to the normal flora?

A: *Clostridium perfringens* and other clostridia involved in gas gangrene (spores are ubiquitous), *C. perfringens* food poisoning; *Clostridium difficile* diarrhea or colitis (usually normal flora but can be contagious), *Clostridium botulinum* food poisoning and infant botulism, and *Clostridium tetani* (normal flora or environmental source).

Q: Anatomically, where are most infections found that are due to *Bacteroides fragilis*?

A: *B. fragilis* is found as normal flora primarily in the intestinal tract, especially in the colon. Therefore most infections are below the diaphragm. However, hematogenous spread can result in infections above the diaphragm.

Q: What is the genus of gram-negative anaerobic rods that have "pointed" ends?

A: *Fusobacterium*. They are part of the gastrointestinal normal flora and can be involved in soft tissue infections. "Fusiform" bacteria are spindle shaped.

Q: The observation of sulfur granules in the exudate from a soft tissue infection suggests what organism is involved?

A: *Actinomyces israelii*. This anaerobe is found throughout the gastrointestinal tract and the female genital tract. It can cause soft tissue infections at any of these sites.

Q: What two genera of anaerobes are sometimes referred to as the black-pigmented anaerobes?

A: *Prevotella* and *Porphyromonas*. These two genera are gram-negative anaerobes and at one time were classified as *Bacteroides*. They are sometimes referred to as the black-pigmented anaerobes because their colonies turn nearly black after prolonged growth on blood agar plates.

Q: What genus of anaerobic bacteria is important in the development of dental caries (tooth decay)?

A: *Actinomyces*, especially *A. viscosus*. Dental caries is a mixed infection and includes other bacteria such as *Streptococcus mutans* (a viridans strep) and *Lactobacillus*.

Q: Within 18 hours after eating a hot beef sandwich smothered with beef gravy, four members of a family of six developed a watery diarrhea accompanied by abdominal cramping. Fever and vomiting were notably absent. Within a day, they felt much better as the symptoms began to resolve. What is the most likely cause of this illness?

A: This is most likely a case of *Clostridium perfringens* food poisoning, most of which are meat associated. "Perfringens" food poisoning is rapid onset and short duration.

Q: From the previous question, what bacterial product causes the symptoms of the food poisoning?

A: An enterotoxin that is produced by the bacteria

Q: From the previous question, where, anatomically, is the toxin produced?

A: In the small intestine, primarily in the jejunum and ileum. Clostridial food poisoning is due to the ingestion of millions of bacteria (about 10^8) that grow in the contaminated food, usually high protein. The bacteria produce the enterotoxin as they sporulate in the small intestine.

Q: *Staphylococcus aureus* is not an obligate anaerobe, but it also causes a food poisoning that is rapid onset and short duration. Where does this organism secrete its enterotoxin?

A: In the food. It is ingested as a preformed toxin.

Q: *Bacillus cereus* is an <u>aerobe</u> that causes a food poisoning that is rapid onset and short duration. With what kind of food is it commonly associated?

A: Cooked rice, including reheated fried rice, that is held at warm temperatures is the usual answer, but it can be associated with other foods including meat and macaroni & cheese. One strain primarily causes diarrhea, another strain primarily causes vomiting.

Q: What is the site of infection in tetanus neonatorum?

A: Umbilicus stump. This is an important cause of neonatal death in underdeveloped countries.

Q: List the signs and symptoms of food-borne botulism.

A: Diplopia, dysphagia, and dysphonia, including symmetric descending paralysis are the most important

Q: Where is the toxin produced in food-borne botulism?

A: In the food after the spores germinate. Because of the ingestion of a preformed toxin, this is an intoxication.

Q: How is food-borne botulism diagnosed?

A: Clinical presentation and by detection of the toxin in the food or patient's serum

Q: What is the therapy for food-borne botulism?

A: Supportive therapy, with administration of equine trivalent antitoxin (types A, B, and E, which are the most common botulism serotypes found in the United States). Serum sickness can occur.

Q: How can food-borne botulism be prevented?

A: Proper canning procedures would help. Cooking contaminated food at 100°C for about 10 minutes inactivates the toxin.

Q: List the signs and symptoms of infant botulism.

A: Infants younger than 1 year of age (usually younger than 6 months), lethargy, constipation, droopy eyelids, and "floppy" head due to flaccid paralysis

Q: Where is the toxin produced in infant botulism?

A: Infants ingest the spores of *Clostridium botulinum*, sometimes with the ingestion of honey, but the spores can be in other foods. The spores germinate in the intestinal tract and the growing cells produce the toxin that is absorbed by the intestinal mucosal cells. The clostridia grow in the infant because there is little competition with the immature normal flora diversity of the infant.

Q: How is infant botulism diagnosed?

A: Clinical presentation

Q: What is the therapy for infant botulism?

A: Supportive, including respiratory support and intravenous or nasogastric feeding if needed. Most infants recover without further intervention. Infant botulism has a low mortality rate. Antibiotics do not halt the progression of the disease

Q: What are the major virulence factors of *Clostridium difficile*?

A: *C. difficile* produces two exotoxins, toxin A and toxin B. Toxin A is an enterotoxin and causes the watery diarrhea. Toxin B is an exotoxin and is related to the pseudomembrane formation.

Q: How is antibiotic-associated pseudomembranous colitis diagnosed?

A: Demonstration of either toxin A or B (pathogenic strains generally produce both) and the demonstration of the yellowish-white pseudomembrane by sigmoidoscopy. Demonstration of the presence of the organisms is not diagnostic because it is commonly part of the normal colonic flora of non-ill individuals, and some strains are nontoxigenic.

Q: What is the most toxic substance known?

A: Botulism toxin. Purified botulinum toxin the size of one grain of table salt could kill 60 people. The toxin is about 100 times more toxic by the inhalational route than by the gastrointestinal route. Of the seven known serotypes of botulinum toxin, serotypes A, B, E, and F are the most common causes of human disease.

Q: As a bioterrorist agent, the Centers for Disease Control and Prevention classify the botulism toxin in what category (A, B, or C)?

A: Category A, the highest priority

Q: Wound botulism is most often seen in what group of individuals?

A: Drug abusers who inject. Symptoms are similar to food botulism.

Spirochetes

Q: Briefly describe the general structure of spirochetes.

A: These spiral-shaped organisms have a gram-negative type of cell wall structure in that they have a periplasmic space between an outer membrane and the cytoplasmic membrane. Bundles of axial fibrils (analogous to flagella) are located in the periplasmic space and are called endoflagella. The endoflagella have their origins near each end of the cell and are arranged in parallel bundles that wind around the length of the cell.

Q: Are all spirochetes motile?

A: Yes. The endoflagella appear to flex and relax, resulting in a twisting motility of the cells.

Q: What organism causes syphilis?

A: *Treponema pallidum*. Molecular evidence suggests that a subspecies of this organism that causes yaws (non–sexually transmitted) in South America was most likely transported back to Europe by Christopher Columbus and his men. The organism became more virulent and developed into a strain of *T. pallidum* that caused a sexually transmitted disease, syphilis. This strain was then transported back to the New World by other Europeans.

Q: What bacteriological medium is used for the primary isolation of *T. pallidum*?

A: None. This organism cannot be grown on artificial media. Research labs grow the bacteria in rabbits.

Q: List the stages of syphilis in humans.

A: There are three stages: primary, secondary, and tertiary.

Q: Briefly describe primary syphilis.

A: Primary syphilis is characterized by the appearance within 3–4 weeks after infection of a single chancre (hard-based ulcer with a sharp edge) that spontaneously heals in 1–2 months.

Q: Are chancres of syphilis painful or painless?

A: Painless

Q: Are the chancres infectious or noninfectious?

A: Infectious

Q: Briefly describe secondary syphilis.

A: Within 1–2 months after the hard chancre of primary syphilis disappears, the papules of secondary syphilis appear on the skin and mucous membranes due to systemic spread of the treponemes. A maculopapular rash can occur on the palms and soles.

Q: Are the papules of secondary syphilis infectious or noninfectious?

A: Infectious

Q: What are the papular syphilitic lesions near the genital area in secondary syphilis called?

A: Condyloma latum. Condylomata acuminata are genital warts (venereal warts) due to papillomaviruses. Genital warts are frequently cauliflower shaped.

Q: What results from infection by *T. pallidum* of hair follicles?

A: Alopecia (hair loss)

Q: Briefly describe tertiary syphilis, including a description of the lesions.

A: Tertiary syphilis can occur 5–40 years after the symptoms of secondary syphilis have abated. It is characterized by noninfectious granulomatous lesions called gummas. Virtually any organ or tissue can be affected, including skin, bones, internal organs (liver, etc.), nervous system (neurosyphilis; brain, meninges, eyes and ears [sensitivity to bright light and loud noises]), and cardiovascular (aneurysms, etc.).

Q: When patients with syphilis are given penicillin, they may have systemic effects such as a transient high fever. Why?

A: *Treponema pallidum*, which has a gram-negative structure, is very susceptible to penicillin. Penicillin causes rapid death and lysis of the bacteria, resulting in the release of endotoxin into the circulatory system. The sudden increase in endotoxin concentration is responsible for the systemic effects.

Q: With respect to the previous question, what is the name of this reaction to the endotoxin?

A: Jarisch-Herxheimer (pronounced yah'rish herks'hī-m̃r) reaction. It is usually not life threatening. The Jarisch-Herxheimer reaction can occur with treatment of other infections such as relapsing fever and Lyme disease.

Q: What is the drug of choice for syphilis?

A: Penicillin G

Q: What microscopic procedures can be used for the diagnosis of syphilis?

A: The gold standard for the diagnosis is the observation of living motile *T. pallidum* from primary or secondary lesions by dark-field microscopy. Fluorescent antibody staining can also be used for specific identification of the organism.

Q: What serological methods are used for the diagnosis of syphilis? Discuss their specificity.

A: Nontreponemal tests such as the RPR and VDRL (you do not need to know what the letters stand for) are screening tests that rely on the nonspecific precipitation of cardiolipin. Antibodies to cardiolipin cross-react with surface antigens on *T. pallidum*. Because of the low specificity, all positive tests must be verified with confirmatory treponemal tests such as FTA-ABS, MHA-TP, and TP-PA (you do not need to know what these letters stand for either). Notice that all of these tests have a "T" (for *Treponema*). Remembering this helps to learn these, because they are specific for detecting *Treponema* infection. These tests are more expensive than the screening tests but detect *T. pallidum*–specific antibodies (thus they confirm the presence of *T. pallidum*–specific antibodies in patients' blood). Recently, a recombinant enzyme linked immunoassay, or EIA, became available as a screening procedure for detection of syphilis-specific antibodies. For a positive EIA screen, the Centers for Disease Control and Prevention has recommended the use of the RPR test in combination with RPR quantitation as the confirmatory method. This approach reduces the total cost of the evaluation because it reduces the use of the more expensive FTA testing in most cases where the EIA and RPR are in agreement.

Q: List some conditions that can give false positives in the nontreponemal tests for syphilis.

A: Systemic lupus erythematosus (up to 10%); some viral infections such as viral hepatitis, mononucleosis, and HIV disease; pregnancy; and malaria are some of the conditions, but there are also others.

Q: A baby born to a mother who had syphilis was diagnosed with the same disease soon after birth. How did the neonate get infected?

A: In utero. Congenital syphilis can lead to stillbirth or serious neonatal abnormalities, but infected neonates are usually asymptomatic at birth, with symptoms developing later if not treated. For those infants that survive, the disease can be classified as early congenital if symptoms manifest before the age of 2 (similar to secondary syphilis in adults) or late congenital if the symptoms occur after the age of 2 (primarily neurological).

Q: What is the major route of transmission to humans of *Leptospira* in the United States?

A: Food or water contaminated with animal urine, especially from dogs

Q: What major organ is affected in leptospirosis?

A: Kidney. Severe manifestation is called Weil disease.

Q: What is the cause of Lyme disease?

A: *Borrelia burgdorferi*

Q: How is Lyme disease transmitted to humans?

A: The bite of the black-legged *Ixodes* tick. Infection rates are therefore most common in warm months when the ticks are active (May to August).

Q: What serves as the animal reservoir for *B. burgdorferi*?

A: The white-footed mouse and white-tailed deer. The *Ixodes* tick serves as the vector that passes the organism between the two hosts and (accidentally) to humans.

Q: What areas of the United States have the highest incidence of Lyme disease?

A: East coast from northern Virginia to southern Maine and the upper mid-west (Wisconsin and Minnesota). Note that Lyme disease can be diagnosed in any state because the symptoms may take up to one month to manifest.

Therefore vacationers to endemic areas might not be symptomatic until after they return to their home (nonendemic) states. The states considered as endemic are Connecticut, Delaware, Maryland, Massachusetts, Minnesota, New Jersey, New York, Pennsylvania, Rhode Island, and Wisconsin.

Q: Describe the hallmark of Lyme disease.

A: A skin rash (erythema migrans) commonly resembling a target or "bulls eye" centered over the tick bite is usually the first sign of infection (70–80% of patients, but some suggest it is seen in as few as 30%). The center is usually clear (but there are many different manifestations) and is surrounded by the expanding annular red rash. The rash usually occurs within 3–30 days after the tick bite. Some patients do not develop the rash.

Q: Describe the progression of untreated Lyme disease.

A: Stage one is characterized by the noninfectious skin rash as described above, frequently accompanied with malaise, fever, chills, headache, and so on. The rash usually fades after 3–4 weeks. In stage two, which begins several weeks to several months later, neurological (e.g., meningitis or encephalitis; ~2%) or cardiac (e.g., arterioventricular block; ≤1%) manifestations can occur. Stage three symptoms occur 5 months to 2 years after stage one. Stage three is characterized by arthritis (~30%), especially in the knees, but may include neurological symptoms resembling neurosyphilis. The arthritis resolves in most patients over the next 2 years, however, it may become chronic.

Q: How is Lyme disease diagnosed?

A: Clinical manifestations (a ≥5 cm diameter erythema migrans by itself within 3–30 days of a tick bite is diagnostic) and serology (demonstration of specific IgG or IgM). A positive enzyme-linked immunosorbent assay or indirect fluorescent antibody followed by a confirmatory Western blot is recommended. On a standardized exam question, a history of traveling in an endemic area and/or a tick bite will probably be mentioned.

Q: What is the treatment for Lyme disease?

A: Doxycycline, amoxicillin, or cefuroxime axetil (expensive)

Q: List some Lyme disease coinfections.

A: Babesiosis (*Babesia microti/divergens*) and human granulocytic anaplasmosis (previously human granulocytic ehrlichiosis, due to *Anaplasma phagocytophilum*)

Q: How is Lyme disease prevented?

A: Using DEET-containing insect repellents, self-examination for ticks, avoiding areas populated with *Ixodes* ticks, wearing long pants and long-sleeved shirts in tick-infested areas, and removing ticks within 24 hours. There is no vaccine.

Q: What is STARI?

A: **S**outhern **t**ick-**a**ssociated **r**ash **i**llness. The illness is due to *Borrelia lonestari* and is found primarily in the southeastern states in the U.S. It is spread by the lone star tick (*Amblyomma americanum*). The rash is similar to the erythema migrans of Lyme disease.

Q: What organism causes tick-borne relapsing fever in the United States?

A: *Borrelia hermsii*. It is spread by soft-bodied ticks (*Ornithodoros hermsi*).

Q: What is the geographical distribution of tick-borne relapsing fever in the United States?

A: West of the Mississippi River, primarily from California, Arizona, and New Mexico north to Idaho and Washington.

Q: What is the geographical distribution of louse-borne relapsing fever?

A: Africa and South America

Q: What is the etiological agent of louse-borne relapsing fever?

A: *Borrelia recurrentis*

Q: Why does relapsing fever relapse?

A: Antigenic variation of outer membrane proteins allows the organism to evade immune surveillance. During bacteremic stages, the organism in the blood has a slightly different antigenic structure than before.

Q: What is tabes dorsalis?

A: It is a condition that results from the demyelination of the dorsal roots and posterior columns in the spinal cord, resulting in loss of senses. Tabes (pronounced tāy bēz) dorsalis is manifested as severe wide-based gait, leg ataxia, postural instability, pain, and paresthesias. It is most commonly associated with neurosyphilis but can have other causes such as spinal cord injury.

Atypical Pathogenic Bacteria: *Mycoplasma, Rickettsia, Ehrlichia, Anaplasma, Chlamydia,* and *Chlamydophila*

Q: What is the difference between the genus *Chlamydia* and the genus *Chlamydophila*?
A: The only significant difference is in DNA homology. Species of *Chlamydophila* were previously classified as species of *Chlamydia*. The biology of both is the same (intracellular growth, etc.). *Chlamydophila* (formerly *Chlamydia*) *pneumoniae* and *Chlamydophila* (formerly *Chlamydia*) *psittaci* are the primary species you need to know. Both genera are in the family Chlamydiaceae. Within the genus *Chlamydia*, you need only be familiar with *Chlamydia trachomatis*.

Q: What species of mycoplasma is responsible for pneumonia?
A: *Mycoplasma pneumoniae*

Q: Describe the signs and symptoms of mycoplasma pneumonia.
A: The infection is most common in teenagers and young adults (15–25 years of age) who present with a persistent nonproductive cough. Because it is a mild infection (that is why it is called walking pneumonia), patients frequently delay seeking medical care. Headache (sometimes severe), malaise, and low-grade fever (37.8–38.9°C) are also common. Rales may be present. Chest x-ray can show evidence of bronchopneumonia. This is an "atypical" pneumonia because of a lack of alveolar exudate that is seen in typical pneumonia, especially that due to *Streptococcus pneumoniae*.

Q: How is *Mycoplasma pneumoniae* transmitted?
A: Aerosol droplets. Infection can become epidemic in some situations such as college dormitories, military barracks, and schoolrooms.

Q: What antibiotics are ineffective for the treatment of *Mycoplasma pneumoniae*?
A: Cell wall active antibiotics, because none of the mycoplasmas have a peptidoglycan cell wall (or any other kind of cell wall).

Q: What are the primary drugs for therapy of *M. pneumoniae* infection?
A: Primary drugs of choice are erythromycin, a tetracycline, azithromycin, and clarithromycin. A fluoroquinolone is a good alternative.

Q: What other bacteria can cause primary atypical (walking) pneumonia?
A: *Chlamydophila pneumoniae, Chlamydophila psittaci, Legionella pneumophila* (Pontiac fever), and *Coxiella burnetii* (Q fever)

Q: What other infections can involve *M. pneumoniae*?
A: Pharyngitis, rhinitis, rhinosinusitis, bullous myringitis, and bronchitis are some.

Q: Mycoplasmas contain what substance in their cytoplasmic membranes that is lacking in other bacteria?
A: Cholesterol, which they obtain from the host or substrate on which they are growing

Q: Describe the colony morphology of the mycoplasmas.
A: "Fried egg" appearance. Mycoplasmas are slow growing. After prolonged incubation (at least one week) on special media the characteristic very small colonies appear. Most laboratories do not use culture for diagnosis of infection.

Q: How is *M. pneumoniae* infection diagnosed?

A: Often, the diagnosis is based on clinical signs, symptoms, x-ray results, and so on. Atypical pneumonia can be due to other organisms as well. Serology, such as complement fixation or enzyme-linked immunosorbent assay (demonstration of fourfold increase in titer between acute and convalescent sera), cold agglutinin titers (IgM; there are false positives and false negatives with this test), and immunofluorescence, can be useful. Complement fixation or titers for specific IgG or IgM are probably used most often and are usually accurate. *M. pneumoniae* can be cultured from sputum or a throat swab, but that can take one or more weeks and is not too sensitive. Polymerase chain reaction and DNA probes are in development.

Q: What other species of mycoplasma are involved in human infections?

A: *Mycoplasma hominis* is involved in genitourinary tract infections such as postpartum and postabortal fever, pelvic inflammatory disease, and pyelonephritis. It can be sexually transmitted. Nongenitourinary infections include septic arthritis, endocarditis, and lower respiratory and posttransplantation infections. *Mycoplasma genitalium* primarily causes sexually transmitted diseases (STDs) such as nongonococcal urethritis in men and similar genitourinary infections in women as *M. hominis*. *Ureaplasma urealyticum* is a mycoplasma that causes similar STDs as those listed above. There are other species that may have a disease association, but you do not need to know any of them.

Trivia: *M. hominis* has the smallest cell and genome of any known bacterium.

Q: What do rickettsia and chlamydia have in common?

A: They are both obligate intracellular parasites and, as with viruses, must reproduce within eukaryotic cells. Obviously, bacteria in the genus *Chlamydophila* (formerly *Chlamydia*) are also obligate intracellular parasites.

Q: List three genera of rickettsia-like organisms.

A: *Rickettsia, Coxiella,* and *Ehrlichia. Anaplasma* (formerly *Ehrlichia*) is a fourth. All are obligate intracellular parasites.

Q: What causes Q fever?

A: *Coxiella burnetii*

Q: Where is *C. burnetii* normally found?

A: It has worldwide distribution (but apparently excluding New Zealand). In endemic areas Q fever (Q for query) has an animal reservoir (sheep [especially], cattle, goats, cats, dogs, wild rodents, ticks [Australia], etc.). Because it is not a reportable disease, not much is known about its epidemiology or incidence of infection in the United States. It is likely underdiagnosed.

Q: Infection due to *C. burnetii* is usually by what route?

A: Inhalation. The organism can survive long periods (a year or more) in harsh conditions, similar to bacterial endospores. *C. burnetii* has a predilection for the placenta. Soil becomes contaminated when animals give birth. It is most often occupationally acquired (abattoir workers, farmers, etc.).

Q: How does Q fever present?

A: Usually, it is not severe and presents as a mild walking pneumonia with symptoms similar to mycoplasma pneumonia (headache, fever, dry cough, etc.). Severe manifestations, such as endocarditis, can have a high mortality rate.

Q: What organism causes Rocky Mountain spotted fever (RMSF)?

A: *Rickettsia rickettsii*

Q: What geographical area of the United States has the highest incidence of RMSF?

A: Southeastern and southern states, from about Maryland south to Georgia, then west through Tennessee and Arkansas to Oklahoma, but it can be found in other states. RMSF was named after the Rocky Mountains because of early cases diagnosed in Montana and later studies performed at the Rocky Mountain Laboratory in Hamilton, Montana.

Q: How does RMSF present?

A: The classical symptom triad for rickettsial diseases, including RMSF, is fever, headache, and rash, but other symptoms, such as chills and myalgias, can exist. The petechial rash spreads toward the trunk from the arms and feet. The palms and soles may have a rash. In some cases the rash is absent. The other classical "triad" to be aware of for RMSF is fever, rash, and history of exposure to ticks.

Q: How is infection with *Rickettsia rickettsii* usually acquired?

A: It is acquired by the bite of ticks: *Dermacentor andersoni* (wood tick), *Dermacentor variabilis* (dog tick), and *Amblyomma americanum* (lone star tick, particularly in Texas and Louisiana). Symptoms occur about one week after the tick bite (3–14 days). The disease incidence parallels when the ticks are most active (April–September).

Q: How is RMSF diagnosed?

A: It is based on the symptoms, especially sudden onset of fever, headache, rash, and a history of tick bite. The rash may be absent, and the patient might not be aware of a recent tick bite. Indirect fluorescent antibody testing is also available.

Q: What is the therapy for RMSF?

A: Doxycycline is the drug of choice. Therapy should be initiated quickly if RMSF is suspected, because the mortality rate increases significantly (pulmonary/renal failure) if treatment is delayed (>20% after five days from onset of symptoms).

Q: What organism causes ehrlichiosis?

A: *Ehrlichia chaffeensis* causes human monocytic ehrlichiosis (HME; infects monocytes). *Anaplasma phagocytophilum* (formerly *Ehrlichia phagocytophila*) causes human granulocytic anaplasmosis (HGA; formerly human granulocytic ehrlichiosis; infects granulocytes).

Q: How does HME present?

A: Fever, headache, myalgia, rash, nausea/vomiting, altered mental status, and lymphadenopathy are most common. It sometimes is referred to as spotless RMSF (RMSF without the rash), but ehrlichiosis commonly has a rash, especially in children. It is most likely misdiagnosed or underdiagnosed.

Q: How is infection with *Ehrlichia* (HME) usually acquired?

A: It is acquired by the bite of ticks: HME is spread by *Dermacentor* (dog) and *Amblyomma* (lone star) ticks; HGA is spread by *Ixodes* (deer) ticks (also the primary vectors for Lyme disease) and *Dermacentor* ticks.

Q: What is the geographical distribution of ehrlichiosis?

A: HME is found primarily in southern states (dogs, goats, and deer are common reservoirs). Human granulocytic anaplasmosis has a distribution similar to that of Lyme disease (northeast, upper Midwest, northern California; deer, wild rodents, and elk are common reservoirs). The distribution of both diseases parallels that of their respective tick vectors.

Q: What is the therapy for ehrlichiosis?

A: Doxycycline is the drug of choice. You probably do not need to know how to diagnose the disease because it can be difficult. Immunofluorescence and Western blot are available in some laboratories.

Q: What is the cause of the most common STD in the United States?

A: *Chlamydia trachomatis*

Q: How are chlamydiae grown in culture?

A: All species of *Chlamydia* (and *Chlamydophila*) are obligate intracellular parasites and must be grown in cell culture. None can grow on agar media. The chlamydiae primarily get ATP from the parasitized cells.

Q: Structurally, do the chlamydiae most closely resemble viruses, gram-positive bacteria, gram-negative bacteria, or mycoplasmas?

A: Gram-negative bacteria. They have an outer membrane that contains lipopolysaccharide and an inner cytoplasmic membrane.

Q: Intracellular chlamydiae can best be viewed with what stain?

A: Because they have a gram-negative structure, the Gram stain first comes to mind. However, the Gram stain is not useful for these bacteria. They can best be visualized intracellularly with other stains such as Giemsa or an immunofluorescent stain.

Q: What are the two developmental forms of chlamydiae called?

A: Elementary bodies and reticulate bodies

Q: From the previous question, which form is infectious?

A: Elementary bodies

Q: Describe the infectious cycle of chlamydiae.

A: The infectious elementary bodies enter cells by phagocytosis. Reorganization occurs to form the reticulate bodies within the phagosomes. The reticulate body divides by binary fission until it contains new elementary bodies. Cell lysis releases elementary bodies that can infect new cells.

Q: Name the three primary pathogenic chlamydiae.

A: *Chlamydia trachomatis, Chlamydophila* (formerly *Chlamydia*) *psittaci*, and *Chlamydophila* (formerly *Chlamydia*) *pneumoniae*. All are in the family Chlamydiaceae.

Q: What are the major infections of these three chlamydiae?

A: *Chlamydophila psittaci*: psittacosis. *Chlamydophila pneumoniae*: atypical pneumonia, bronchitis, pharyngitis, and rhinosinusitis. There is evidence that this organism may also be involved in cardiovascular disease. *Chlamydia trachomatis*: trachoma, conjunctivitis (infant and adult), infant pneumonia, nongonococcal urethritis, and lymphogranuloma venereum.

Q: What is psittacosis and how does infection occur?

A: Psittacosis, sometimes called parrot fever, is a pneumonia caused by the inhalation of dried bird feces. Many species of birds, including poultry, can carry the organism. It is also called ornithosis.

Q: In what age group is *Chlamydophila pneumoniae* most commonly a pathogen and how is it acquired?

A: Young adults (but anyone older than age 5 years) are infected by inhalation of aerosolized respiratory droplets.

Q: How does *C. pneumoniae* pneumonia present?

A: Slow onset after infection, often preceded by pharyngitis. It presents as a mild chronic cough with malaise. Chlamydial pneumonia resembles other walking pneumonias.

Q: How is the diagnosis of *C. pneumoniae* pneumonia made?

A: Serologically by demonstration of a fourfold increase in IgG or IgM. Some labs may grow the bacteria in cell culture for the observation of glycogen-filled inclusions after staining with iodine. Polymerase chain reaction can also be used.

Q: What is the therapy for *C. pneumoniae* pneumonia?

A: First drugs of choice include a tetracycline, erythromycin, clarithromycin, or azithromycin. The most important thing to remember is that cell wall active antimicrobials, such as the beta-lactam antibiotics, do not work.

Q: What is trachoma?

A: It is an eye infection caused by *Chlamydia trachomatis*. It can lead to scarring of the cornea and blindness. Not all strains of *C. trachomatis* can cause trachoma. The strains are separated into different types called biovars and serovars. Some of these biovars and serovars cause only trachoma, whereas others are the agents of other infections such as conjunctivitis and urethritis or lymphogranuloma venereum.

Q: As an STD, how does *C. trachomatis* present in males?

A: As with gonorrhea, men are usually symptomatic, primarily with urethritis (penile discharge, burning on urination, etc.).

Q: As an STD, how does *C. trachomatis* present in females?

A: Again, as with gonorrhea, most women are asymptomatic (symptomatic presents as thick yellow vaginal discharge, itching, and burning on urination). It is the number one cause of pelvic inflammatory disease and can result in sterility and ectopic pregnancy. Other manifestations include urethritis, cervicitis, endometritis, and salpingitis. Cervical erosion may be present.

Q: How are genital infections of *C. trachomatis* diagnosed?

A: Clinical presentation and laboratory diagnosis. Laboratory tests include culture, direct immunofluorescence, enzyme immunoassay, nucleic acid hybridization tests, and nucleic acid amplification tests.

Q: How are genital infections of *C. trachomatis* treated?

A: Either azithromycin or doxycycline is the drug of choice for urethritis and cervicitis. On a standardized exam, a tetracycline will probably be the correct answer for chlamydia questions. In the United States patients with other STDs, such as gonorrhea, are usually treated for chlamydia because it is so common.

Q: Genital infections due to *C. trachomatis* can lead to what syndrome?

A: Reiter syndrome

Q: How does Reiter syndrome present?

A: It presents as a triad of arthritis, urethritis (or cervicitis), and conjunctivitis (or uveitis). Reiter syndrome (aka reactive arthritis) is an autoimmune disease due to cross-reacting antibodies to surface antigens of *C. trachomatis* and several other bacteria. There is a strong association of Reiter syndrome (a spondyloarthropathy) with persons who have the major histocompatibility complex class I antigen HLA-B27.

Q: What other organisms can cause Reiter syndrome?

A: The gastrointestinal pathogens *Campylobacter, Shigella, Salmonella,* and *Yersinia.*

Q: What species of chlamydia causes inclusion conjunctivitis in infants and how is it contracted?

A: Neonates become infected with *Chlamydia trachomatis* as they pass through the birth canal.

CHAPTER 12 | Virology

Q: What is the protein coat of a virus called?
A: Capsid. The nucleic acid/capsid complex is called the nucleocapsid.

Q: What are capsomeres?
A: They are the polypeptide subunits of the capsid.

Q: What is the name of the process by which virus subunits are assembled into infectious viruses?
A: Maturation

Q: How many animal viruses contain both DNA and RNA?
A: None. All viruses contain either DNA or RNA, but not both.

Q: How are animal viruses detected by culture (what do you look for)?
A: Some viruses are detected in cell culture by the effect they have on the cells (e.g., rounding up, cell fusion, death, sloughing off, etc.). This is called cytopathic effect. Others can be detected by immunological or molecular techniques (e.g., polymerase chain reaction [PCR]).

Q: What is a plaque assay?
A: A plaque assay is actually not used for virus detection but rather for virus quantitation. It can be used for some animal viruses, including human, but is particularly useful for the quantitation of bacteriophages of bacteria. To perform the assay (described here for animal viruses), a solution of viruses is serially diluted. A measured volume from each dilution tube is added to susceptible cells that are then incubated. After incubation, the tissue culture cells are stained with a dye that is taken up only by living cells. Dead cells do not take up the dye. A plaque is a clear spot in the cell monolayer where a single virus initially infected a single cell and as it reproduced and spread surrounding cells were infected and died. By counting the number of plaques at various dilutions, it is possible to calculate the total number of infectious virions before dilution.

Q: List the DNA virus families.
A: "PPPPAHH" = Parvo-, Polyoma-, Papilloma-, Pox-, Adeno-, Hepadna-, Herpesvirus. If you reverse the order, you can use the mnemonic "HHAPPPPy" (and ignore the "y") if this is easier for you. Hepadnavirus is easy because it has "dna" in the middle. If you learn the seven families of DNA viruses, it makes it easier to identify the RNA virus families if given a choice. This is fortunate because there are 14 RNA families and only 7 DNA families (learn the 7 first, not the 14). Therefore, by default, if the virus is not in one of the seven DNA families that you learned, it must be in one of the RNA families. Similarly, if the virus family is given and it is not one of the DNA families you learned, it must be an RNA virus family. Note: One additional family of RNA viruses, Birnaviridae, may infect humans. There is some evidence that it is involved in some human psychiatric diseases such as schizophrenia, mood disorders, and autism, but this has not been proven. If this family is included, there are 15 RNA families. There are additional DNA and RNA virus families, but they do not contain human pathogens.

Q: To see if the tip on learning virus families works, try this question: What kind of nucleic acid do the Togaviruses have? What about Paramyxoviruses?
A: RNA for both. It worked!

Q: Of the DNA virus families, which are double stranded (ds) and which are single stranded (ss)?
A: All are dsDNA except the Parvoviruses, which contain ssDNA.

Q: Of the RNA virus families, which are dsRNA and which are ssRNA?
A: From the previous question, most DNA viruses are double stranded. The reverse is true for the RNA viruses. All are single stranded except one family, the Reoviruses, which includes both rotavirus (gastroenteritis) and

reovirus (Colorado tick fever). There are several others that are not significant human pathogens. Another odd characteristic about Reoviruses is that their genome is segmented.

Q: What other viruses have a segmented genome?

A: Segmented genomes are only found among the RNA viruses. The segments are not identical to each other. The other viruses, besides the Reoviruses, that have segmented genomes are the Arenaviruses (e.g., Lassa fever), Orthomyxoviruses (e.g., influenza), and Bunyaviruses (e.g., Hantavirus cardiopulmonary syndrome, Rift Valley fever).

Q: There is one RNA virus family that has two strands of ssRNA in which both strands are identical. Which family?

A: Retroviruses (technically, the family Retroviridae). Although retroviruses have two strands of RNA, they are not thought of as being segmented because both strands are identical. Instead, we think of them as being diploid (they are the only diploid viruses).

Q: There is only one known parvovirus that causes human disease. What is it and what does it cause?

A: Parvovirus B19 causes fifth disease (erythema infectiosum). It can also be a cause of aplastic anemia, arthritis, and abortion.

Q: What role do the papillomaviruses play in human disease?

A: Human papilloma viruses (HPVs) are a cause of warts (skin, genital, etc.) and cervical cancer.

Q: Can HPV infection be prevented?

A: Yes. A vaccine was approved in the United States by the Food and Drug Administration in 2006. This is the first vaccine developed to prevent cervical cancer, precancerous genital lesions, and genital warts due to HPV types 6, 11, 16, and 18. The vaccine is approved for use in females ages 9–26 years. The risk of male-to-female transmission of HPV infection can be reduced with male condom use.

Q: What are the human herpesviruses (HHVs) and their primary diseases? (Note: This is best answered by making a table with the paper in landscape orientation. If you leave enough room, you can add other columns to include signs/symptoms, lab diagnosis, treatment, and so forth. This technique works well for other pathogens as well.)

A: Herpes simplex virus 1 (HSV-1; cold sores), HSV-2 (genital herpes), varicella-zoster virus (VZV; chicken pox, shingles), Epstein-Barr virus (EBV; infectious mononucleosis, Burkitt lymphoma, Hodgkin disease, oral hairy leukoplakia), cytomegalovirus (CMV; immunocompromised, fetus), HHV-6 (roseola [aka sixth disease]), HHV-7 (roseola [not as important as HHV-6]), and HHV-8 (aka Kaposi sarcoma–associated herpesvirus). (Note: This is the short list; there are other HHV disease associations.)

Q: What role do the polyoma viruses play in disease?

A: There are two human polyoma viruses, the JC virus and the BK virus. They generally do not cause serious human disease. Most people are infected with JC virus but are asymptomatic. Immunosuppression can cause reactivation and have serious consequences such as fatal progressive multifocal leukoencephalopathy. Similarly, BK virus infection is common, but most people are asymptomatic carriers. BK virus can cause mild respiratory disease and sometimes more serious disease in the immunocompromised. They are DNA tumor viruses but do not cause tumors in their natural hosts. If they appear on a standardized exam, they will probably be one of the wrong answers.

Q: What type of agent causes Creutzfeldt-Jakob (pronounced croits-felt yă-cub) disease?

A: A prion. Note: Recently, the role of prions in transmissible spongiform encephalopathies has been challenged with the suggestion that an as yet unknown virus may be the cause. Until that is proven, it will be assumed that prions are the cause.

Q: What are prions?

A: Prions are infectious pathogens composed of proteins. They are unique in that they contain no nucleic acid (DNA or RNA). They appear to be modified proteins of normal cellular genes. The prion protein has the unique ability to affect the conformation of the normal protein (normal cellular proteinaceous particle, or PrP) so that it becomes a prion (i.e., infectious form of PrP and is designated as PrP^{Sc} [Sc from scrapie]).

Q: What is the difference between Creutzfeldt-Jakob disease and variant Creutzfeldt-Jakob disease?

A: The naturally occurring disease in humans is called (classic) Creutzfeldt-Jakob disease and is late onset (median age at onset in the United States is about 68). If the disease is acquired by eating meat from infected animals, it is called variant Creutzfeldt-Jakob disease (median age at onset is 26). The disease in cattle is commonly referred to as mad cow disease.

Q: A virus isolated from a patient was found to be inactivated by ether. What can be said about the structure of this virus?

A: It possesses an envelope.

Q: How do enveloped viruses acquire their envelope?

A: During maturation, as they bud through either the plasma membrane or the nuclear membrane of the host. From the previous question, ether is a lipid solvent; therefore it causes irreversible damage to the lipid containing membrane that constitutes the envelope.

Q: What viruses are naked (no envelope) and which ones are enveloped?

A: This is easier to answer if you just remember those that are naked because most viruses do have an envelope. If you learn the naked ones, all the others are enveloped. Among the DNA viruses (seven families), four are naked: parvoviruses (fifth disease), adenoviruses, papillomaviruses, and polyomaviruses. Remember these families as PAPP (the patient must be naked to get a papp [sic] smear). Of the remaining three DNA virus families (Hepadnaviruses [hepatitis B], Herpesviruses, and Poxviruses; all enveloped), the Poxvirus family, which also starts with "p," is unique among all viruses in that it has a complex morphology. Among the RNA viruses, only 3 of the 14 families are naked: Picornavirus (polioviruses, coxsackieviruses, echoviruses, and enteroviruses), Calicivirus (norovirus, hepatitis E), and Reovirus (rotaviruses, Colorado tick fever). One way to remember these families is with the acronym PCR (in the PCR reaction, "naked" nucleic acid is used). You still need to memorize these three, because there are other RNA viruses that begin with "p," "c," and "r" that are enveloped.

Q: Assume that a single cell is infected with two different strains of the same DNA virus. One of these viruses has a gene called *Xa* that codes for a specific protein on its surface and another gene called *Ya* that codes for a different protein. The other virus has a genome that is identical to the first, but instead of genes *Xa* and *Ya* it has *Xb* and *Yb* that code for slightly different proteins. After replication of the viruses and subsequent cell lysis, normal XaYa and XbYb virions are detected as well as some XaYb and XbYa virions. What accounts for the appearance of the XaYb and the XbYa genotypes?

A: Most likely, there was a recombination event that occurred during replication. This resulted in the replacement of one gene (e.g., *Ya*) with another nearly identical gene (e.g., *Yb*). It is easier to understand this type of question if you draw a picture.

Q: Let's try another genetics question (do not forget to draw a picture with this question). A cell is infected with two different strains of the same virus. The first strain has capsid "A" (capsid composed of protein A) and genotype "a" (written as Aa). The second virus has capsid "B" and genotype "b" (Bb). After replication of the viruses and subsequent cell lysis, normal Aa and Bb virions are detected as well as some Ab (type A capsid and type b genome) and Ba virions (type B capsid and type a genome). What accounts for the appearance of the Ab and the Ba virions?

A: This is an example of phenotypic mixing in which the genome of one virus is packaged inside the capsid of another.

Q: What is a defective virus?

A: It is a virus that has a mutation, deletion, or some other genetic aberration that results in inactivation of a function critical for replication. Defective viruses can infect cells, but they do not normally replicate.

Q: How can a defective virus replicate?

A: If a cell infected with a defective virus is coinfected with another virus (a helper virus) that has a functional gene that can complement the defective gene on the defective virus, then the helper virus can supply the missing component (e.g., enzyme) necessary for replication. It is possible for two defective viruses in the same cell to replicate if they can complement each other's defective genes. In this situation, each virus would be a helper virus relative to the other.

Q: Are defective viruses important in human medicine?

A: You knew the answer was "yes" or the question would not have been asked (see next question).

Q: List some examples of defective viruses that infect humans.

A: Some oncogenic viruses are defective. They originate in cells where during the replication cycle they replace normal virus genes with human genomic genes. As with other defective viruses, they replicate only if the cell is coinfected with a helper virus. Another example of a defective virus is hepatitis D virus (HDV). HDV requires a helper virus (hepatitis B virus) for replication.

Q: Which of the following types of enzymes can inactivate prions: nucleases, lipases, or proteases?

A: Proteases, because prions are made of protein. Prions, however, are more resistant to proteases than other proteins.

Q: What is meant by viral load?

A: Number and concentration of viral particles, in a body fluid, usually the blood plasma. Generally, a high viral load correlates with higher morbidity. Monitoring viral load is useful in determining the effectiveness of therapy.

Q: How is viral load measured?

A: There are several methods. It is probably most important to know about measuring HIV viral load, but it can be used with other viruses such as hepatitis B and C, cytomegalovirus, and Epstein-Barr. A given number of viral particles always yield the same quantity of nucleic acid. There are three FDA approved methods for measuring HIV RNA. In one test, reverse transcription of the RNA into DNA is followed by PCR. The PCR product can then be quantified, such as with an enzyme-linked immunosorbent assay (ELISA). Nucleic acid sequence–based amplification is similar to reverse transcriptase-PCR. The branched DNA (bDNA) test amplifies a signal (the amount of light emitted by chemicals that associate with the viral RNA). Viral load for DNA viruses, such as EBV and CMV, can be tested by real-time PCR. (Note: Both reverse transcription PCR and real-time PCR are abbreviated as RT-PCR. Real-time PCR is also known as kinetic PCR, qPCR, qRT-PCR, and RT-qPCR [q = quantitative].)

Q: A 2-year-old patient presents with a hoarse, barking, "seal-like" cough that is worse at night. What is the most likely cause?

A: Parainfluenza viruses (there are several serotypes) most frequently cause croup (also known as laryngotracheobronchitis). It is usually preceded by an upper respiratory tract infection. A barking, seal-like cough is a characteristic of croup. It is most often seen in the winter months and in children 3 months to 5 years of age. Parainfluenza viruses can cause other upper respiratory tract infections (colds, sore throats) and lower respiratory tract infections (pneumonia, bronchitis, bronchiolitis) as well as diseases that do not involve the respiratory tract, such as febrile illness and infections of the central nervous system (CNS).

Q: What other respiratory viruses cause croup?

A: Respiratory syncytial virus (RSV), adenoviruses (there are six subgroups [A–F] and various serotypes [e.g., 1, 2, 3, etc.] within the subgroups), measles virus, and influenza viruses A and B. Croup can also have a bacterial etiology (*Haemophilus influenzae*).

Q: What are coronaviruses and what kinds of infections do they cause?

A: They are small enveloped, nonsegmented, long single-stranded, (+) sense RNA viruses. They can infect many species of vertebrates. The human coronaviruses primarily are respiratory pathogens. They are the second most common cause of the common cold. They can also cause gastrointestinal infections, especially in infants.

Q: What severe coronavirus infection originated in China in 2002?

A: Severe acute respiratory syndrome, known as SARS. The virus is annotated as SARS-CoV. As of this writing, there are no active cases of SARS anywhere in the world.

Q: Can breast milk from HIV-positive mothers infect their babies?

A: Yes. Breast milk from slightly less than half of infected mothers may contain infectious virions. Therefore in developed countries, HIV-positive women are advised to use formula instead of breast-feeding.

Q: Name a common virus that causes a bloody diarrhea.

A: Trick question! Viruses cause watery diarrheas. If a patient has a bloody diarrhea, consider bacteria such as *Shigella*, *Campylobacter*, *Salmonella*, some strains of *Escherichia coli*, or *Yersinia enterocolitica*.

Q: List some causes of viral gastroenteritis.

A: Some of the more common are norovirus (number one), rotavirus (number one in children), adenovirus, astrovirus, coronavirus, Coxsackievirus, and echovirus. Infections are usually acute onset and watery.

Q: List the most common causes of viral pneumonia.

A: Influenza virus, parainfluenza virus (especially infants and children), adenovirus (especially children), and RSV (most common cause of bronchiolitis and pneumonia in infants and children <1 year of age)

Q: Is there a vaccine currently available for RSV?

A: No, but one might be available soon.

Q: What is the cause of fifth disease?

A: Parvovirus B19

Q: Describe the clinical presentation of fifth disease.

A: It normally affects children, who acquire it via the respiratory route. It often has a sudden onset and presents with a bright red macular rash on the cheeks, giving the child a "slapped cheek" appearance. It usually is self-limiting. Fifth disease is also called erythema infectiosum. The rash of scarlet fever is raised in comparison.

Q: How are influenza viruses classified (what family)?

A: They are Orthomyxoviruses.

Q: What kind of nucleic acid is in the Orthomyxoviruses?

A: They contain single-stranded linear RNA that is segmented. Influenza A has eight segments.

Q: Where in the infected cell does the influenza virus replicate?

A: In the nucleus. This is unusual, because most RNA viruses replicate in the cytoplasm. Retroviruses also replicate in the nucleus.

Q: Among influenza viruses, what is antigenic shift and drift?

A: As mentioned, influenza virus is an orthomyxovirus containing eight separate pieces of ssRNA. Antigenic shifts are major changes due to the reassortment of the segments of viral RNA during mixed infections (frequently occurring in animals, such as pigs). With these mixed infections, progeny viruses have the complete complement of all the required RNA segments, but with antigenic shift the segments come from each of the different infecting influenza viruses. These new influenza variants (subtypes) have an antigenically distinct hemagglutinin–neuraminidase combination different from the original infecting viruses. These new variants are the sources for the "new" influenza viruses that frequently emerge to which the general population is not immune. Antigenic drift is due to mutations (often point mutations) that cause minor changes in the neuraminidase and hemagglutinin on the surface of a subtype of the virus.

Q: With influenza A virus, is a change from H_2N_2 to H_3N_2 an antigenic shift or drift?

A: Shift, because there has been a major change in the hemagglutinin

Q: What does the H_5N_1 influenza serotype suggest?

A: Avian influenza

Q: Is there a vaccine for avian flu?

A: Yes. The FDA approved an inactivated, monovalent avian influenza virus vaccine for the active immunization of adults aged 18 through 64 years who are at increased risk for exposure to the H_5N_1 subtype. An adequate immune response was attained in 45% of subjects. Although antibody levels were not optimal in the remaining patients, current scientific information suggests that disease severity and influenza-related hospitalization and death may still be reduced.

Q: "Owl's eye" intranuclear inclusions can be observed in cells infected with what virus?

A: CMV

Q: Name the five primary childhood exanthems.

A: Measles (rubeola), rubella (German or 3-day measles), fifth disease (erythema infectiosum), roseola (roseola infantum [sixth disease], exanthema subitum), and chickenpox

Q: All families of RNA viruses (there are 14) have ssRNA except one. Which family has dsRNA?

A: Reoviridae (reo: respiratory enteric orphan). These viruses have a segmented genome containing from 10 to 12 segments, depending on the virus.

Q: What Reoviridae are important human pathogens?

A: The most important is rotavirus. Colorado tick fever virus (genus = coltivirus) is another but usually is not on standardized exams.

Q: What is the most common cause of viral gastroenteritis?

A: In the United States and most developed countries, the most common cause of viral diarrhea (adults + children) is norovirus (family Caliciviridae). The rotaviruses are the primary agents of infectious diarrhea in infants and children (>500,000 deaths/year worldwide). Other diarrhea-causing viruses include the astroviruses (usually children), adenoviruses, and sapoviruses (same family as norovirus).

Q: How does norovirus infection present?

A: The incubation period is usually 1–2 days but can range from 8 to 56 hours and presents with vomiting and/or diarrhea, usually with fever, abdominal cramps, or nausea (at least two of the three).

Q: How is norovirus infection confirmed?

A: PCR. Viral gastroenteritis is usually not laboratory confirmed in an individual unless there is a suspected large-scale outbreak, such as on a cruise ship.

Q: How does rotavirus diarrhea present?

A: After about a two-day incubation period, mild to severe illness can occur. In developed countries, occasional cases of fatal gastroenteritis due to dehydration and electrolyte imbalance may occur. Symptoms include fever, vomiting, and abdominal pain with a nonbloody, watery diarrhea that commonly lasts from three to eight days. Severe dehydration and death in children is more common in emerging nations. Rotavirus infections are most common in the winter months in temperate climates.

Q: How is rotavirus infection diagnosed?

A: Rapid antigen detection in stool specimens. Strains may be further characterized by enzyme immunoassay or reverse transcriptase PCR, but such testing is not commonly done. The virions can be observed in stool using an electron microscope (they resemble wheels [rota]).

Q: What is the treatment for rotavirus infection?

A: For the diarrhea, rehydration, including electrolytes. In developing countries the mortality rate is high, due to the dehydration that accompanies the disease. Several live oral vaccines for use in preventing rotavirus gastroenteritis in infants are available.

Q: How is poliovirus transmitted and where does it replicate (anatomically, not where in infected cells)?

A: Fecal–oral route (it is an enterovirus). It replicates in the oropharynx and intestinal tract before hematogenous spread to the CNS. Replication in the motor neurons of the anterior horn leads to cell death and the concomitant paralysis.

Q: How does poliovirus infection <u>usually</u> present?

A: Most infections are asymptomatic.

Q: How do most cases of symptomatic poliovirus infection present?

A: Aseptic meningitis. Paralytic disease is not common.

Q: What is meant by aseptic meningitis?

A: Meningitis without routine bacterial isolation from cerebrospinal fluid

Q: What is the most common cause of aseptic meningitis?

A: Viruses, especially the enteroviruses such as coxsackie and echoviruses. Note that some atypical bacteria such as mycoplasmas, *Borrelia*, and *Treponema* (among others) can cause aseptic meningitis.

Q: Why is the Sabin live vaccine preferred to the Salk killed vaccine in endemic areas?

A: There is a better immune response because the growth of the attenuated live virus in the Sabin vaccine results in a prolonged stimulation of the immune system.

Q: In the United States and many other geographical regions of the world where the incidence of polio infection is rare, an inactivated poliovirus vaccine is preferred. Why?

A: There is a remote possibility of reversion to a virulent strain if the live vaccine is used, resulting in vaccine-associated polio. Currently in the United States, the inactivated polio vaccine is recommended at 2 and 4 months, again from about 6 to 18 months, and the last vaccination at 4–6 years of age.

Q: What is meant by a latent viral infection?

A: Infection with no overt symptoms. The latent virus may be reactivated later to cause overt disease.

Q: Name some viruses that can cause latent infections.

A: CMV, EBV, HHV-7, HSV, VZV, and HIV are some.

Q: Heterophile-positive mononucleosis is most likely due to what virus?

A: EBV. Heterophile antibodies cross-react with commonly shared antigenic epitopes found in widely divergent species.

Q: Heterophile-negative mononucleosis is most likely due to what virus?

A: CMV

Q: Besides mononucleosis, what other diseases have been associated with EBV?

A: Most Burkitt lymphomas and nasopharyngeal carcinomas (squamous cell carcinoma) as well as many B-cell lymphomas

Q: What is the source of new strains of influenza A viruses?

A: Animals such as swine, horses, and chickens

Q: What is the predominant antibody produced after initial infection with viruses?

A: IgM, followed by a gradual increase in IgG

Q: How is the mumps virus spread?

A: Respiratory, via air-borne droplets

Q: How does mumps present?

A: Mumps is generally a childhood disease, presenting after 2–3 weeks with fever and swelling of one or both parotid glands. Complications include (among others) orchitis (in men), meningitis, and encephalitis.

Q: Is the mumps virus vaccine dead or live?

A: Live (attenuated). It is one component of the MMR vaccine.

Q: The genome of the mumps virus consists of single-stranded negative-sense RNA and the rubella virus consists of single-stranded positive-sense RNA. Which RNA is infectious and why?

A: The rubella RNA is infectious because it can function as messenger RNA since it is positive sense. Negative-sense RNA requires viral RNA-dependent RNA polymerase to synthesize complementary positive-sense RNA. With these viruses, the complementary positive-sense RNA serves as a template for the synthesis of additional negative-sense RNA that is packaged within the virus capsid. Messenger RNAs are also synthesized using complementary positive-sense RNA as the template.

Q: Compare the clinical presentation of measles (rubeola) and rubella (German measles).

A: **Measles:** cold-like symptoms followed by high fever and Koplik spots (red spots with tiny white/gray centers found on oral mucosa). A maculopapular rash (flat, red area covered with small papules) follows, spreading from the head to the extremities. Photophobia is common.

Rubella: low-grade fever often accompanying faint diffuse macular rash without the papules (also spreads from head to extremities). Generally, rubella is a mild disease but can cause congenital deformities (congenital rubella syndrome), especially if infection occurs in the first trimester during pregnancy.

Q: What causes roseola?

A: Primarily, HHV-6. There are two genetic variants: HHV-6A (no known disease associations) and HHV-6B (a mononucleosis syndrome, encephalitis, pneumonitis, others). Evidence has shown that one particular type of epilepsy (mesial temporal lobe epilepsy) is associated with HHV-6B infection.

Q: How does roseola present?

A: Most cases occur before 12 months of age (maternal antibodies are protective in the first few months) with an incubation period of about 5–15 days. Roseola, most often due to HHV-6, presents with the rapid onset of a high fever (39–41°C) that can last up to five days, followed by a red macular rash resembling the rash of rubella, but generally lasts no more than two days.

Q: Which hepatitis viruses are transmitted in contaminated food or water?

A: Hepatitis A and E. Hepatitis E is found in some developing countries and is unlikely to appear on a standardized exam. The few cases of hepatitis E infection in the United States are usually acquired by traveling in endemic areas.

Q: A mother brings her 5-year-old son to the clinic because he was just bitten by the neighbor's dog. She is frantic because she fears he will now develop rabies. What should you tell her about the incidence of rabies to calm her fears?

A: The Centers for Disease Control and Prevention declared the United States canine-rabies free in 2007. The incidence of human rabies in the United States is 0–3 cases a year. The likelihood that her son is infected is extremely low.

Q: How should the child from the previous question be treated?

A: The wound should be cleansed with soap and water. The patient should begin the immunization series (five injections of the rabies vaccine over a one-month period) and should be given human rabies immune globulin (or equine antiserum if human is not available). Even though the chances of developing rabies in this case are extremely low, rabies is a fatal infection and the above treatment should be initiated.

Q: From the previous question, why is the use of soap important?
A: The rabies virus is enveloped. Soap, detergents, or other lipid solvents can inactivate the virus.

Q: In the United States, what animals are most likely to be infected with the rabies virus?
A: Bats, skunks, and raccoons. In the U.S. most human cases are from bats. Dogs are the usual carriers in developing countries.

Q: After the symptoms of rabies virus infection occur, what is the most common outcome?
A: Most symptomatic cases are fatal. There are six documented survivals in the world (all included therapy).

Q: Describe the intracorporeal (within the body) travel of the rabies virus.
A: The virus is spread centripetally up the peripheral nerves into the central nervous system. The incubation period for rabies is usually 30–60 days with a range of 10 days to one year. The route of transmission usually occurs via infected secretions, saliva, or by infected tissue. Stages of the disease include upper respiratory tract infection symptomatology, followed by encephalitis. The brainstem is affected last.

Q: What virus is associated with Kaposi sarcoma?
A: HHV-8 (aka Kaposi sarcoma–associated herpesvirus)

Q: What cells can serve as a target for HIV?
A: The primary target is the CD4+ helper T lymphocyte, but it also infects other CD4+ cells such as macrophages, monocytes, dendritic cells, and microglial cells.

Q: Is CD4 the only receptor protein involved in HIV infection?
A: No. The primary receptor for HIV entry is CD4, but an additional coreceptor is required. The chemokine receptor CCR5 can serve as a coreceptor for HIV. It is found (along with the CD4 receptor) on T cells (T helper, memory, activated), natural killer cells, monocytes, macrophages, immature dendritic cells, and Langerhans cells. Because the CD4 receptor is also found on these cells, all can be infected by HIV. Other cells, such as basophils, have CCR5 but not CD4, so they cannot be infected with HIV. A small minority of people lack CCR5 and are resistant to HIV infection. Some HIV strains can use the chemokine receptor CXCR4 (fusin) on T lymphocytes as a coreceptor. About half of HIV-infected individuals who have been treated have an HIV strain that can use the CXCR4 receptor.

Q: What kind of nucleic acid is found in prions?
A: None. Prions are composed of protein.

Q: What does Sin Nombre virus cause?
A: Sin Nombre virus ("unnamed virus" in Spanish) causes hantavirus pulmonary syndrome (hantavirus cardiopulmonary syndrome, hantavirus disease). Sin Nombre virus is the most common strain of hantavirus in the United States. Others include Bayou and Black Creek Canal viruses. Most cases in the United States are in the western portion of the country. It has a case fatality rate of about 50%. It is an RNA virus in the family Bunyaviridae.

Q: How is hantavirus infection acquired?
A: It is acquired by inhalation of aerosolized saliva and excreta in areas frequented by the deer mouse (western United States) or white-footed mouse (eastern United States). Rats and voles can also serve as reservoirs, and infection is dependent on the virus strain and geographical location.

Q: What is the treatment for hantavirus pulmonary syndrome?
A: Supportive only

Q: What is the vector for dengue fever?
A: *Aedes* mosquito, most commonly *A. aegypti*

Q: How is the virus that causes dengue fever classified?
A: It is a flavivirus (enveloped, positive-sense, ssRNA). It can also be classified as an arbovirus, but arboviruses refer to the mode of transmission (arthropod-borne) of viruses, not their structure. There are four serotypes of the dengue fever virus. Other families of viruses contain arboviruses.

Q: How does dengue fever present?

A: Dengue is characterized by acute onset of high fever, rash, bone pain ("breakbone fever," although there is no increase in bone fractures), headache, and myalgia. Severe disease presents as a hemorrhagic fever with death due to hypotension and shock.

Q: What is chikungunya?

A: It is an infection similar to dengue but is more severe. The arthritis and crippling arthralgia are clinically distinct. It is caused by an alphavirus (a Togavirus [ssRNA]) and is spread by *Aedes* mosquitos, primarily in Africa and Southeast Asia. The disease is a major health threat in these locations. It would be unusual to have a question about this disease on an exam.

Q: Oral hairy leukoplakia can occur in association with what infection?

A: AIDS

Q: What is the cause of oral hairy leukoplakia?

A: EBV

Q: What infections does VZV cause?

A: Chickenpox (varicella) and shingles. Infection, especially primary infection in adults, can result in varicella pneumonia.

Q: Should adults be vaccinated against varicella?

A: Adults at any age who have no evidence of varicella immunity should be vaccinated.

Q: What is shingles?

A: Shingles is also called herpes zoster (VZV is classified as a herpes virus) and is due to reactivation of latent VZV from the dorsal root ganglia. Anyone who has had chickenpox is at risk for developing shingles. It occurs most frequently in older adults, usually after age 50, and in others with a compromised immune system. The painful maculopapular lesions usually occur within a dermatome. It is estimated that one in five people will develop shingles in their lifetime.

Q: Can shingles be prevented?

A: A qualified yes. There is a vaccine available to be given at age 60. It can prevent shingles about half the time. It is most effective for persons younger than age 70, with decreasing effectiveness as people age. The vaccine for shingles is not a substitute for the chickenpox vaccine.

Q: A mother is worried that her 5-year-old will get chickenpox because she was playing with her neighbor, who was diagnosed with chickenpox. The neighbor (also 5 years old) had crusty lesions all over his body. If this was the only day she played with the neighbor, will she develop chickenpox too?

A: No. Chickenpox is only contagious 24–48 hours before the rash breaks out and until the vesicles have crusted over.

Q: What body fluids can contain HIV?

A: HIV is most commonly transmitted via semen, vaginal secretions, blood, and breast milk, but it has been found in saliva, urine, cerebrospinal fluid, tears, alveolar fluid, synovial fluid, transplanted tissue, and amniotic fluid. There has not been documentation of infection from casual contact.

Q: How quickly do patients infected with HIV progress to AIDS?

A: Five to 10% develop symptoms within three years of seroconversion (seroconversion is usually within three months of infection); 95% develop AIDS within 15 years. The mean is about 8–10 years for adults and two years for children less than 5 years old.

Q: When should antiretroviral treatment begin in an asymptomatic patient with HIV?

A: Before the CD4+ T-cell count falls below 350 cells/mm^3

Q: Drug resistance is common with bacterial infections. Is drug resistance of concern with HIV?

A: Yes. Drug resistance testing is recommended for HIV-1 before treatment is initiated. Many strains of the virus are multiply drug resistant. Studies have shown that patients whose physicians have access to drug resistance data respond better to therapy compared with patients whose HIV-1 strains were not tested for resistance.

Q: What is the recommended initial treatment for HIV infection?

A: The U.S. Department of Health and Human Services Panel on Antiretroviral Guidelines for Adults and Adolescents recommends two nucleoside/nucleotide reverse transcriptase inhibitors (e.g., tenofovir and emtricitabine or zidovudine and lamivudine, etc.) and either a nonnucleoside reverse transcriptase inhibitor (e.g., efavirenz, etc.) or a ritonavir-boosted or unboosted protease inhibitor (e.g., atazanavir, fosamprenavir, lopinavir).

Q: A patient was found to be infected with HIV-2 rather than HIV-1. What would you ask the patient about travel history?

A: Ask about travel to West Africa. Although HIV-2 can be found in other parts of the world, it is most common in West African countries. AIDS with HIV-2 is usually slower onset.

Q: What is the recommended treatment for genital herpes?

A: Oral acyclovir, famciclovir, or valacyclovir

Q: Which strain of influenza is most common in adults and which is most common in children?

A: Adults: influenza A. Children: influenza B

Q: Which hepatitis viruses can be transferred by blood?

A: Hepatitis B (HBV), C (HCV), and D (HDV) viruses. This is a hint I presented earlier: List the five hepatitis viruses in order: A, B, C, D, and E. A and E are on the *outside* of the other three viruses. Hepatitis A and E come from *outside* (contaminated food/water), thus fecal–oral. The three remaining viruses are B, C, and D and are on the *inside* of A and E. Blood is on the *inside* of the body; therefore these three (B, C, and D) are blood-borne.

Q: Which hepatitis viruses are enveloped?

A: The three blood-borne, HBV, HCV, and HDV. Use the above hint, but slightly modified: think of them as on the *inside* of an envelope.

Q: Which hepatitis viruses can result in chronic infections?

A: The same three, HBV, HCV, and HDV. Hepatitis A and E are usually acute onset.

Q: Which hepatitis viruses have DNA and which have RNA?

A: All hepatitis viruses have RNA except HBV, which has DNA.

Q: Which hepatitis virus will not replicate unless the host is coinfected with HBV?

A: HDV is defective and requires a helper virus (HBV). It is most common in intravenous drug abusers who are infected with HBV. The HDV envelope contains the HBsAg.

Q: How is HAV infection prevented in someone not yet exposed to the virus?

A: Vaccination. Recovery from infection also confers lifelong immunity.

Q: How is HBV infection prevented in someone not yet exposed to the virus?

A: Vaccination. With approximately 2 billion people infected worldwide and 360 million with chronic HBV infection, including 30% of cirrhosis and 50% of hepatocellular carcinoma, vaccination has a high potential of decreasing morbidity and mortality.

Q: How is HCV infection prevented in someone not yet exposed to the virus?

A: Use a safe blood supply (transfusions) and behavior modification in those at risk for spreading the virus, especially intravenous drug abusers. Other sources include body piercing/tattooing and unprotected sex with partners whose health status is unknown.

Q: How is HDV infection prevented in someone not yet exposed to the virus?

A: HDV is defective and requires an active HBV infection for replication. Thus the HBV vaccine prevents HDV infection.

Q: How is HEV infection prevented in someone not yet exposed to the virus?

A: Use of a safe water supply. HEV infection is rare in the United States and most developed countries with safe water supplies. It is most common in Asia, China, and Mexico.

Q: What is HBsAg?

A: It is a major surface antigen of the envelope of HBV. It is the first marker of HBV infection.

Q: What is HBcAg?

A: The hepatitis B core antigen. Its presence gives rise to the core antibody, anti-HBc.

Q: The clinical laboratory reported a patient was anti-HBcAg (+; IgM) and anti-HBs (−). What can be said about his infection status?

   A: Acute infection; a high anti-HBcAg (**IgM**) titer means high infectivity, whereas a low titer suggests chronic active infection. Anti-**HBs** antibody appears later in the infection and suggests recovery and immunity. Anti-HBs antibody appears during convalescence. Anti-HBcAg (**IgG**) is always detected in HBV infection. A high titer of IgG anti-HBc without IgM anti-HBc suggests a persistent infection.

Q: The clinical laboratory reported a patient was HBsAg (−) and anti-HBs (+). Is this patient infectious?

   A: No. The *presence* of the HBsAg indicates infection; in the above scenario the patient is HBsAg (−), indicating that there is no active infection. The HBsAg is the first indication of infection. Anti-HBs antibody is a marker for immunity. This patient either had a prior infection or has been vaccinated.

Q: IgG HBcAg (+), anti-HBs (−) is indicative of what?

   A: Past infection, not infectious

Q: What is HBeAg?

   A: The hepatitis B envelope antigen. Its presence indicates active viral replication and high patient infectivity.

Q: HBeAg (+) is indicative of what?

   A: Ongoing viral replication in the liver, highly infectious

Q: HBeAg (−) and emergence of anti-HBeAg antibody indicates what?

   A: This so-called seroconversion is an indicator of clinical improvement of the hepatitis infection (reduced virus replication and reduced infectivity).
   Note: Molecular assays (quantitative viral load tests, genotyping assays, drug resistance mutation tests, and core promoter/precore mutation assays) are becoming more widely available for use in diagnosis and management of HBV infection.
   Another Note: In answering viral hepatitis questions on a standardized exam, pay attention to the c, s, and e and if it is the antigen (Ag) or the antibody (anti) that is being referenced.

Q: Which of the hepatitis viruses are associated with hepatocellular carcinoma (hepatoma, liver cancer)?

   A: HBV and HCV

Q: What is the composition of the HBV vaccine?

   A: Hepatitis B surface antigen (HBsAg) is the major immunizing component.

Q: The warty genital lesions (venereal warts) of papillomaviruses are called what?

   A: Condylomata acuminata. They are soft, skin colored, and are frequently cauliflower shaped. Condyloma latum are the papular syphilitic warts in the genital area and are usually softer and smoother than condylomata acuminata. HPV types 6 and 11 are most common causes, but others can be involved.
   Note: A vaccine was approved in the United States by the FDA in 2006 for types 6, 11, 16, and 18. The vaccine is approved for use in females ages 9–26 years. The risk of male-to-female transmission of HPV infection can be reduced with male condom use.

Q: Where is polio endemic?

   A: At the time of publication of this book, Nigeria, India, Afghanistan, and Pakistan

Q: Differentiate between the Sabin and Salk polio vaccines.

   A: The Sabin vaccine is a live attenuated poliovirus. The Salk vaccine is an inactivated (killed) vaccine. A better immune response is attained with the Sabin vaccine because the living virus provides a longer stimulation of the immune system.

Q: The final step of herpes virus replication is budding through a membrane to acquire the envelope. Through which membrane do they bud, cytoplasmic or nuclear?

   A: Nuclear. All other enveloped viruses bud through the cytoplasmic membrane. Herpesviruses are DNA viruses. The other DNA viruses, except the poxviruses (cytoplasm), also replicate in the nucleus, but the others do not bud through the nuclear membrane.

Q: Where do RNA viruses replicate?

   A: Most replicate in the cytoplasm. The exceptions are the orthomyxovirus (influenza) and the retroviruses (HIV), which both replicate in the nucleus.

Q: In general, RNA viruses exhibit high mutation rates. Why?

A: RNA viruses require RNA polymerases for replication. These enzymes typically make mistakes that cannot be corrected (enzymes involved in DNA synthesis of eukaryotic nuclear DNA do have repair capabilities).

Q: Do DNA viruses have high mutation rates?

A: It depends on the virus. Four families (Parvo-, Polyoma-, Papilloma-, and Hepadnaviruses) use host enzymes for replication and therefore mistakes can be repaired. However, the other three DNA virus families (Adeno-, Herpes-, and Poxviruses) code for their own enzymes that are not as efficient at proofreading and repairing.

Q: A 23-year-old female medical student presented to student health with headache, fever, enlarged lymph nodes and spleen, and pharyngitis. There was an increase in atypical white blood cells, and the laboratory reported the detection of heterophile antibodies. What cells are typically infected?

A: Initial infection involves epithelial cells of the oropharynx, but later in the infection B cells are primarily involved.

Q: With respect to the previous question, what virus is involved and what is the disease?

A: EBV/mononucleosis. Remember the connection to heterophile antibodies (EBV causes heterophile-positive mononucleosis).

Q: If B cells are infected with EBV in a laboratory experiment, what happens to the cells?

A: They are immortalized. This means the cells can be serially passed in the laboratory indefinitely. Some cells grown in tissue cultures in the lab, such as primary cells, can only be passed (grown in successive transfers) a limited number of times before they die (they are mortal). Other cells, such as many cancer cells (transformed cells) or some virus infected cells, are immortal.

Q: What are heterophile antibodies?

A: They are antibodies (usually IgM) produced in response to some foreign antigens (i.e., EBV and some trypanosomes) that are nonspecific. They can react not only with the inducing antigen, but also with other antigens such as animal red blood cells. Infection with CMV does not induce the production of heterophile antibodies (i.e., it causes heterophile-negative mononucleosis).

Q: What lab test can be used to detect heterophile antibodies?

A: The most recognizable is the Monospot™ (based on horse red cells), but there are others, such as an enzyme-linked immunosorbent assay.

Q: During EBV infection in mononucleosis, the virus replicates in infected cells. In response, antibodies are made to some of the virus antigens. Detection of these antibodies is another way of detecting infection. In addition, the antibodies are produced in a sequence that indicates the disease progression. List the antibodies (there are three primary ones) or the antigens and indicate at what stage of disease they occur.

A: If you believe you know the answer, compare it with the answer in the stem of the following question. If you need a hint, go to the following question.

Q: There are three primary antibodies produced in response to EBV in mononucleosis: anti-early antigen, anti-viral capsid antigen, and anti-EBV nuclear antigen. In what stage of disease do they appear?

A: Anti-early antigen and anti-viral capsid antigen appear early, whereas anti-EBV nuclear antigen appears late. However, anti-viral capsid antigen (IgM) appears early and anti-viral capsid antigen (IgG) appears late (IgG titers begin to increase as IgM titers decrease).

Q: How does mononucleosis present?

A: Splenomegaly, lymphadenopathy, and exudative pharyngitis are the most important to remember, but fatigue (malaise) and fever are usually present.

Q: What other viruses can cause pharyngitis?

A: Rhinovirus, adenovirus, parainfluenza virus, coxsackievirus, coronavirus, echovirus, herpes simplex virus, and cytomegalovirus

Q: A Pap smear from a 48-year-old woman revealed the presence of koilocytes. What virus is most likely responsible?

A: HPV. HPV serotypes 16, 18, 31, and 33 are most commonly involved in cervical carcinoma. A vaccine was approved in the United States by the FDA in 2006 for HPV types 6, 11, 16, and 18.

Q: Describe the microscopic appearance of koilocytes.

A: They are squamous cells with enlarged, wrinkled, dark ("raisinoid") nuclei surrounded by a halo. Koilocytes can also be observed in HPV infections at other sites, including men or persons with AIDS.

Q: A white 36-year-old man presented with acute onset of a severe headache. He appeared lethargic and confused. Cerebrospinal fluid lymphocytes were elevated. A temporal lesion was seen by functional magnetic resonance imaging of the brain. What virus is most likely the cause?

A: HSV. HSV-1 is the most common cause of sporadic (nonepidemic) encephalitis in adults, usually involving one of the temporal lobes. HSV-2 is more commonly involved in meningitis.

Q: What is the treatment for herpes encephalitis?

A: Acyclovir. Untreated cases have a high mortality rate. Most cases of encephalitis due to other viruses are treated with supportive care only. Because herpes encephalitis is treatable, a definitive diagnosis is important.

Q: How is herpes simplex encephalitis diagnosed?

A: The definitive diagnosis is by cerebral biopsy (culture and PCR). However, because of the associated risk of brain biopsy, the clinical presentation, including the detection of temporal lobe abnormalities (by computed tomography, technetium brain scanning, or gadolinium magnetic resonance imaging) can be highly suggestive of the diagnosis. Serologies are not as important for HSV as for some other viruses (EBV, CMV, HHV-6 [roseola], and mumps). It is more important for you to know the types of tests available (primarily serology, culture, and PCR) for encephalitis diagnosis than it is for herpes simplex virus specifically.

Q: What are the human pox viruses and their primary diseases?

A: The two primary ones are the variola virus (smallpox) and the molluscum contagiosum virus. There are occasional human infections with animal pox viruses (e.g., monkeypox), but humans are not their primary hosts. Note: Chickenpox is not a pox virus.

Q: How are the rashes of smallpox and chickenpox differentiated?

A: The firm deep-seated vesicles or pustules of smallpox are typically all in the same stage of development and are mainly on the face and extremities (centrifugal distribution). The superficial vesicles of chickenpox are commonly described as "dewdrop on rose petal" and are in different stages of development; their distribution is mainly on the trunk and fewest on the extremities. Smallpox has an acute onset of fever (>38.3°C; 101°F) before the rash. Chickenpox has no or mild prodrome.

Q: At what stage of smallpox do patients become noncontagious?

A: After the pustules crust over, form scabs, and all the scabs fall off. This can be up to three weeks from the appearance of the rash.

Q: How should a smallpox patient be handled and what else should be done to prevent others from getting the disease?

A: Immediate isolation of the patient in a type C (contagious) facility with nonshared ventilation systems that exhaust air through a HEPA filter to the outside. All caregivers must use personal protective equipment, which includes an N95 respirator (at a minimum), gown, and gloves. The health department should begin identifying and vaccinating exposed individuals.

Q: It was revealed in the past medical history that a male patient was vaccinated against smallpox in 1958 when he was 13. Is he still immune to smallpox?

A: Probably. Humoral and cell-mediated immunity against smallpox is retained and can be detected clinically for at least 20 years after vaccination. It appears that even people inoculated in the 1950s have maintained a robust immune response.

Q: What is the smallpox vaccine composed of and how is the vaccine administered?

A: The vaccine is composed of live vaccinia virus (derived from the cowpox virus). A bifurcated needle is used in vaccination. It is preferred to vaccinate over the deltoid muscle, but the triceps muscle can be used. After cleansing of the site with soap and water, the bifurcated needle is inserted vertically into the vaccine vial, resulting in a droplet of vaccine adhering between the needle prongs. The droplet contains the recommended dosage of vaccine. For vaccination, the needle is held perpendicular to the skin and 15 punctures are rapidly made with strokes vigorous enough to allow a trace of blood to appear.

Q: What are the adverse reactions to smallpox vaccination?

A: Inadvertent inoculation (autoinoculation to another site), generalized vaccinia (systemic spread), eczema vaccinatum (extensive skin involvement), progressive vaccinia, and postvaccinial encephalitis (~40% fatality rate).

Q: Persons with what conditions should not be vaccinated against smallpox?

A: Persons with previous allergic reaction to smallpox vaccine or any of the vaccine's components; anyone who is immunocompromised, including those with moderate or severe acute illness or heart disease; children younger than age 12 months; women who are pregnant (because of risk to fetus) or breast-feeding; and those with latex allergy (vaccine vial stopper contains rubber)

Q: Eczema vaccinatum is a rare, but serious, complication of smallpox vaccination that most commonly occurs in persons who have skin conditions such as eczema (atopic dermatitis). How should it be treated?

A: Administration of vaccinia immune globulin. Note: Erythema multiforme is a hypersensitivity reaction that may occur, but it has causes other than smallpox vaccination and usually resolves on its own with no special treatment.

Q: Would a 5-year-old be more likely to have antibodies to HSV-1 or HSV-2?

A: HSV-1 infection is more common in children than HSV-2 infection. HSV-2 can be a cause of neonatal herpes.

Q: How does HSV-1 infection present in children?

A: Although infection with HSV-1 can be asymptomatic, acute disease presents as gingivostomatitis (ulcerating vesicles in and around the mouth that heal within two weeks).

Q: What infections are associated with HSV-2?

A: Primarily genital and neonatal herpes, but HSV-1 can cause some of the same infections, especially in adults (~20% of genital herpes is due to HSV-1, ~80% due to HSV-2).

Q: What is the source of the herpes virus in *recurrent* cold sores and *recurrent* genital herpes?

A: Cold sores (herpes labialis or fever blisters) are usually due to a reactivation of HSV-1 that has been dormant in the trigeminal ganglia. Genital herpes (usually HSV-2) remains latent in the sacral ganglia until reactivated.

Q: Describe the appearance of herpes labialis.

A: As either primary or reactivation lesions, they appear as grouped vesicles on an erythematous base. They eventually rupture to produce serous or hemorrhagic crusts. They can resemble the vesicles of impetigo.

Q: What is the Tzanck test?

A: It is a microscopic test for HSV (oral, genital, etc.) or VZV infection. It is also called a Tzanck prep or smear. The base of a blister is scraped and smeared on a glass microscope slide. The slide is stained and observed microscopically. Epidermal multinucleated giant cells (keratinocytes) are characteristic of herpes virus infections.

Q: The cerebrospinal fluid of a patient was sent to the laboratory for evaluation. The results suggested viral meningitis. In general terms, what were the results (what are the typical cerebrospinal fluid laboratory findings in viral meningitis and how do they differ from bacterial or fungal meningitis)?

A: There was probably an increase in white blood cells, as there would be with bacterial and fungal meningitis (especially PMNs), with normal protein and glucose. In bacterial and fungal meningitis, the glucose would be low and the protein elevated. The most important value for differentiating viral from bacterial or fungal is the low glucose.

Q: RNA viruses can be either positive stranded or negative stranded. How does each type replicate? This is easier to answer (and remember) if you draw a picture.

A: The RNA of positive-stranded RNA viruses is equivalent to messenger RNA (mRNA) and therefore can be immediately translated into proteins. Some of these proteins are enzymes that function to make negative-stranded RNA. These negative strands can then be used as a template to make more positive-stranded RNA that can be packaged into the viral head of the newly synthesized viruses. Because the positive strands are equivalent to mRNA, they also serve as templates for protein translation (e.g., enzymes necessary for replication and structural proteins). The retroviruses also have positive-stranded RNA, but their replication cycle is unique.

The RNA of negative-stranded RNA viruses cannot function as mRNA. Instead, it must be converted into positive-stranded RNA (i.e., mRNA). Negative-stranded RNA viruses contain their own enzyme (RNA-dependent RNA polymerase) in the virus particle that can convert the negative strand into a positive strand. This positive-stranded RNA can then serve as a template for both proteins (structural and enzymes) and negative-stranded RNA. The negative-stranded RNA is finally packaged, along with RNA-dependent RNA polymerase, to form mature viruses.

Q: List some human retroviruses and their disease associations.

A: The four human retroviruses you need to know are as follows: (1) Human T-cell leukemia virus (HTLV)-1 is associated with asymptomatic infection (most common), adult T-cell leukemia, and tropical spastic paraparesis, a neurological disease; (2) HTLV-2 has no proven disease associations (although some suspected); and (3) HIV-1 (worldwide) and (4) HIV-2 (primarily West Africa) cause AIDS. All infect CD4 bearing cells.

Q: In general terms, describe the replication of retroviruses (as before, drawing pictures helps to remember).

A: After infection and uncoating in the cytoplasm of the cell, the positive-stranded RNA of the virus is used as a template to synthesize a negative-stranded DNA (this is now a hybrid of double-stranded nucleic acid composed of one strand of positive-stranded RNA and one strand of negative-stranded DNA). Synthesis of this negative-stranded DNA from the virus RNA is mediated by a reverse transcriptase that was carried into the cell within the infecting virus. The reverse transcriptase then removes the positive-stranded RNA and replaces it with positive-stranded DNA, resulting in dsDNA of viral origin. During the replication of the RNA into DNA, the end sequences of the RNA are duplicated to form the long terminal repeats (LTRs) that are at each end of the DNA. The dsDNA then enters the cell nucleus where it is integrated into the host chromosome as a provirus (proviral DNA). During normal transcription of host DNA into mRNA, the provirus is transcribed into retrovirus genomic RNA (to be packaged into structural proteins to make mature virions) or viral mRNA (for making viral structural and regulatory proteins).

Q: What is the function of the LTRs in retroviruses?

A: The LTRs are used for integration of the virus into the host genome and they contain promoter and enhancer sequences required for transcription.

Q: Besides LTRs, all retroviruses have at least three other genes. What are they and what are their functions?

A: The *gag* (group-specific antigen) gene codes for structural proteins, including the capsid. The *pol* (**polymerase**) gene codes for reverse transcriptase, protease, and integrase. The *env* (**envelope**) gene codes for surface glycoproteins.

Q: In HIV, what is the designation of the major products of *gag* and *env* and what are their functions?

A: The **gag gene** codes for a precursor protein called p55, which along with other viral and cellular proteins triggers the budding of the viral particle from the surface of an infected cell. After budding, p55 is cleaved by the virally encoded protease, during the process of viral maturation, into four smaller proteins called MA (matrix protein or p17), CA (capsid protein or p24), NC (nucleocapsid or p9), and p6. Do not worry too much about p9 and p6. Most MA molecules help to stabilize the virion by attaching to the inner surface of the envelope of the virion; however, some binds to the integrase. CA is a structural protein found in the core of viral particles.

The *env* gene product is a glycoprotein called gp160. It is subsequently cleaved by a cellular protease to form two smaller envelope glycoproteins, gp120 (also called SU [surface]) and gp 41 (also called TM [transmembrane]). The gp120 is located on the surface of the virion and is attached noncovalently to gp 41 that is part of the virus envelope. The gp120 protein binds to the CD4 receptors when cells are initially infected.

Q: A previous question asked about the genes that all retroviruses possess. In addition to those genes (*gag, pol, env,* and the LTRs), HIV-1 has additional genes. What are some of the more prominent genes and what is their function (i.e., what is the function of their translated protein)?

A: The *tat, rev,* and *nef* genes code for the regulatory proteins Tat, Rev, and Nef. Tat (transcriptional transactivating protein) enhances the transcription of mRNA. It is an RNA binding protein that binds to the transactivation response element, which is a short-stem loop structure that is located at the 5' terminus of HIV RNA. Rev is an RNA binding protein that acts to facilitate the export of viral RNA from the nucleus to the cytoplasm. Nef, another regulatory protein, has multiple activities including down regulation of the cell surface expression of CD4 and the stimulation of HIV infectivity.

Q: An HIV-positive male presents with white lesions on the edges of the tongue. He most likely has an infection due to what virus?

A: EBV. He has oral hairy leukoplakia.

Q: What is the treatment for oral hairy leukoplakia?

A: Acyclovir (which is also the treatment for HSV in HIV-positive patients) or topical podophyllin resin. Newer antivirals such as valacyclovir and famciclovir are also effective.

Q: What is HAART?

A: Highly active antiretroviral therapy, which refers to the combination of three to four drugs for treating HIV infection. Examples of HAART are two nucleoside reverse transcriptase inhibitors (NRTI) + one protease inhibitor, two NRTIs + one non-nucleoside reverse transcriptase inhibitor (NNRTI), or two NRTIs + ritonavir (a protease inhibitor) + another protease inhibitor. There are other alternative combinations. The FDA recently approved a single pill combining efavirenz + tenofovir + emtricitabine for once-daily administration. Tenofovir is a nucleotide reverse transcriptase inhibitor (NtRTI).

Q: List some NRTIs, nucleotide reverse transcriptase inhibitors, NNRTIs, protease inhibitors, and fusion inhibitors.

A: NRTIs (listed alphabetically): abacavir, didanosine, emtricitabine, lamivudine, stavudine, zalcitabine, and zidovudine

Nucleotide reverse transcriptase inhibitors (NtRTIs): There is only one FDA-approved at this writing, tenofovir disoproxil fumarate.

NNRTI: delavirdine, efavirenz, and nevirapine

Protease inhibitors: amprenavir, atazanavir, darunavir, fosamprenavir, indinavir, lopinavir (available in United States only in combination with ritonavir), nelfinavir, ritonavir, saquinavir and tipranavir

Fusion inhibitors: There is only one, enfuvirtide. The mechanisms of action of these were discussed in Chapter 3. Make sure you know the mechanism of action of each group.

Q: List some infections caused by adenoviruses.

A: Most infections are in children, but adults are also at risk. About half of all infections are asymptomatic. Most adenovirus diseases affect the respiratory tract, such as tracheitis and mild pharyngitis. Other diseases include the common cold, pneumonia, croup, bronchitis, and conjunctivitis. They can also cause gastrointestinal disease, including diarrhea and intussusception in infants. Various serotypes (over 50) are more common in each of these diseases than are others. For example, adenovirus serotype 7 has been shown to cause fulminant bronchitis and pneumonia in infants. A variant of adenovirus serotype 14 has recently emerged as a cause of serious and sometimes life-threatening acute respiratory disease in all ages, including healthy young adults. Immunocompromised patients are at higher risk for pneumonia, hepatitis, encephalitis, pancreatitis, nephritis, and hemorrhagic cystitis.

Q: What is the most significant structure that allows for the easy identification of adenoviruses by electron microscopy?

A: At each of the 12 pentons, there are fibers (topped with a small knob) that extend out from the virion.

Q: What is the function of the fibers of the adenoviruses?

A: They function in attachment of the virus to cells and they have hemagglutination activity. The terminal knob of the fiber is critical for binding to cellular receptors.

Q: What are some infectious causes of acute bronchitis?

A: Viral etiologies are most common, including influenza A and B, parainfluenza, RSV, coronaviruses, adenoviruses, and rhinoviruses. Bacterial causes (less than 10% of cases) include *Mycoplasma pneumoniae*, *Chlamydophila pneumoniae*, and *Bordetella pertussis*. Because the microbial cause of acute bronchitis is rarely identified in clinical practice, routine cultures for identification of the etiologic agent are not recommended.

Q: What are the signs and symptoms of acute bronchitis?

A: It typically presents with a cough, with or without phlegm production, which lasts for up to three weeks, and a normal chest radiograph (radiology to rule out other diseases; patient has no signs of pneumonia). Diagnosis is commonly made after ruling out other causes (pneumonia, common cold, acute asthma, or exacerbation of chronic obstructive pulmonary disease). The infection is usually self-limiting.

Q: What is the role of antibiotics in the treatment of acute bronchitis?

A: They should not be used for routine treatment, unless a bacterial cause (e.g., *Bordetella pertussis*) is highly suspected.

Q: How is chickenpox spread?

A: Respiratory droplets, direct contact with vesicles, indirect contact with fomites

Q: What is the recommended age for the first hepatitis B vaccine?

A: At birth, before hospital discharge. It is the only vaccine given at this age.

Q: The influenza virus has two important glycoproteins on the surface of its envelope. What are they and what are their functions?

A: The attachment of virions to host cell receptors is mediated by the hemagglutinin (H antigen). The neuraminidase (N antigen) is important after infection for release of the virus from infected cells.

Q: With respect to the last question, the host cell receptors are composed of what?

A: Sialic acid on epithelial cells

Q: What are the signs and symptoms of influenza virus infection?

A: Acute onset of any or all of the following: nonproductive cough, sore throat, fever (usually high), myalgia, headache, malaise, rhinitis

Q: What drugs are approved in the United States for preventing or treating influenza? Under what conditions are they used and what is their mode of action?

A: Two drugs, zanamivir (inhaled) and oseltamivir (oral), are 70–90% effective for prophylaxis against the flu if taken within 48 hours of exposure. Both are neuraminidase inhibitors of influenza types A and B. They are sialic acid analogs and prevent release of viruses from infected cells. They can also decrease the severity and duration of the disease if taken within 48 hours after becoming symptomatic. Because of the emergence of resistance, amantadine and rimantadine are no longer recommended.

Q: What therapy is used to cure genital warts due to HPV?

A: There is no cure.

Q: What is the treatment for removal of genital warts?

A: Trichloroacetic acid, podophyllin, and liquid nitrogen (cryotherapy) are most commonly used and have mixed succuss. Imiquimod cream and podofilox (gel or solution) can be self-applied and are just as effective. The HPV vaccine is not used for therapy.

Q: How is the laboratory diagnosis of influenza virus infection made?

A: Classically, it involves isolation of the virus from respiratory secretions. The virus in cell cultures is identified on characteristics such as cytopathic effect, inhibition of hemagglutination or hemadsorption, and serological identification of viral antigens. Rapid diagnostic tests are now available that do not rely on isolation in cell cultures. These include screening immunoassays for detection of viral antigens in secretions or enzyme-based tests for the detection of the neuraminidase. A negative screen does not rule out the presence of influenza virus.

Q: What is the drug of choice for VZV infection?

A: Acyclovir for varicella; acyclovir, valacyclovir, or famciclovir for herpes zoster. Foscarnet or cidofovir can be used for VSV (and HSV) if it is resistant to acyclovir.

Q: Ganciclovir is used for what viral infections?

A: CMV, especially retinitis, colitis, and esophagitis in immunocompromised patients (including AIDS). It is most effective by intravenous administration. Valganciclovir is more effective orally than ganciclovir.

Q: When is ribavirin indicated?

A: For patients with chronic hepatitis C, combination therapy with ribavirin and pegylated interferon is the treatment of choice. It can also be used with caution (because of possible adverse side effects) in severe RSV infection in children. It has also been used successfully in other diseases (e.g., Lassa fever and hantavirus infection).

Q: What is the vector for Colorado tick fever?

A: The wood tick, *Dermacentor andersoni*

Q: How does Colorado tick fever present?

A: Most cases are asymptomatic or subacute. Acute cases present with sudden onset of fever, chills, myalgia, headache, and so on. Leukopenia is present and is an important diagnostic aid. The infection is characterized by acute onset of symptoms, a brief asymptomatic period, and then a second febrile phase.

Q: What other infections (nonviral) can be spread by *Dermacentor* ticks?

A: The most important is RMSF. *Dermacentor* variabilis (American dog tick) and *Dermacentor andersoni* (Rocky Mountain wood tick) are the primary vectors for RMSF. *Dermacentor* ticks can also serve as a vector for *Francisella tularensis* (tularemia). Both *Dermacentor* and *Ixodes* ticks can cause tick paralysis (due to a tick salivary neurotoxin, not an infectious agent).

Q: What causes the common cold?

A: A better question might be "what does *not* cause the common cold?" The most common viruses that cause the common cold include rhinoviruses, coronaviruses, adenoviruses, and coxsackieviruses, but cold-like symptoms can also be due to influenza viruses, parainfluenza viruses, and RSV. Mycoplasma infection can also present with symptoms of the common cold but also have lower respiratory tract signs.

Q: How is CMV classified?

A: It is a herpesvirus (enveloped, dsDNA).

Q: Most adults have been infected with CMV. Can a fetus be infected with CMV?

A: Yes. CMV infection is usually asymptomatic in immunocompetent adults, including pregnant females. Once infected, an individual carries the virus for life. Reactivation of the virus may or may not occur but most often is seen in immunocompromised persons. There is a high transmission rate from actively infected mothers to child either in utero or at the time of delivery. CMV is the most common congenital infection and can lead to abortion, stillbirth, neurological manifestations, or birth defects. The fetus will not be infected unless the mother becomes infected during pregnancy and has no prior immunity. Transmission can also occur through breast-feeding.

Q: Who should not receive the chickenpox vaccine? Why?

A: It is a live vaccine and should not be given during pregnancy or to immunocompromised patients because it may cause active disease in the fetus or patient. There are a few more contraindications (e.g., allergies to components, currently ill, etc.), but the live vaccine connection is most important for you to know.

Q: A 6-year-old nonvaccinated boy presents with painful swelling of the parotid glands and pain on swallowing. A viral infection is suspected. What is a common potentially serious complication in adults who become infected with this virus?

A: Orchitis, but sterility is rare. Sterility is more likely if both testes are affected. The boy probably has mumps. After initial infection and localized replication in the pharynx or conjunctiva, viremia can lead to infections in many tissues and organs (e.g., salivary glands, testes, ovaries, pancreas, brain, etc.).

Q: How is the mumps virus classified?

A: It is a paramyxovirus.

Q: What are some other important paramyxoviruses?

A: Measles, RSV, and parainfluenza

Q: What is another name for measles?

A: Rubeola

Q: How is measles spread?

A: Respiratory droplets

Q: What type of vaccine is the measles component of the MMR vaccine?

A: Live (attenuated). MMR (all 3 components), varicella (chickenpox), rotavirus, Sabin polio, influenza (nasal spray), vaccinia (smallpox), zoster (shingles) and yellow fever are the live viral vaccines. Others are in development. The only live bacterial vaccine available in the U.S. is for typhoid fever.

Q: Persistent measles virus infection can lead to a very rare, but serious, neurological disease months to years after the initial infection. What is it called?

A: Subacute sclerosing panencephalitis

Q: Reye syndrome is associated with what viruses?

A: Influenza A and B, VZV (chickenpox), measles, rubella, CMV, and mumps. The precise role of salicylates is not known, but there is a link between Reye syndrome and the use of aspirin and other salicylate-containing medications during virus infections. Because of an intense public awareness program, the incidence is extremely low.

Q: How does Reye syndrome present?

A: Symptoms usually occur in a child (<15 years old) recovering from a recent viral infection and include uncontrollable vomiting and nausea followed by lethargy, irritability, disorientation or confusion, convulsions, and loss of consciousness. There is an increase of cranial pressure and accumulation of fat in the liver and other organs.

Q: Keratitis (inflammation of the cornea) can have multiple etiologies. What is the most common viral cause?

A: Herpes simplex virus

Q: What is the vector for the St. Louis encephalitis virus?

A: The *Culex* mosquito, which is also the vector for Western equine encephalitis. The primary vector for the other major mosquito-borne encephalitides (Eastern equine encephalitis, Venezuela equine encephalitis, and California encephalitis) is *Aedes*. Either mosquito can transmit Venezuela equine encephalitis.

Q: How does St. Louis encephalitis present?

A: Most cases are inapparent or mild. Symptoms appear 7–21 days after a bite from an infected mosquito (usually in the summer or early fall). It can present as a headache with fever and other flu-like symptoms, aseptic meningitis, or encephalitis. The other viral encephalitides have a similar presentation.

Q: List some other viral causes of encephalitis.

A: This is another long list. In the United States the arboviruses (e.g., West Nile, St. Louis, Western equine, and La Crosse encephalitis viruses) are the most common causes. Eastern equine encephalitis is one of the most deadly causes, but it is rare compared with the others. Other viral causes include HSV, EBV, CMV, VZV, HIV, adenovirus, dengue, enteroviruses, rabies, measles, and mumps. You do not need to memorize all the viruses. Just be aware that many viruses can cause encephalitis.

Q: List some viral causes of meningitis.

A: The same viruses from the previous question. Echoviruses (they are enteroviruses) are the number one causes of aseptic (viral) meningitis. Coxsackieviruses are also common.

Q: What is the cause of hand-foot-and-mouth disease?

A: Coxsackie viruses (several serotypes). Vesicles in the oral cavity, hands, and feet characterize the disease. Young children are at highest risk.

Q: What is human metapneumovirus?

A: First described in 2001, human metapneumovirus is a paramyxovirus that causes respiratory tract infections resembling those of RSV (possibly second most common behind RSV in causing lower respiratory tract infections in young children). Diseases include common cold syndrome, bronchiolitis (one of leading causes; RSV is the most common cause), pneumonia, and croup.

Q: Coxsackie viruses cause another infection in children that is characterized by acute onset of fever and sore throat accompanied by vesicles with ulceration in the oral cavity, especially the posterior of the pharynx and soft palate (can affect uvula, tonsils, and tongue). What is it?

A: Herpangina. Coxsackievirus A is the most common cause, but coxsackievirus B, echovirus, or other enteroviruses can be causes. Herpes viruses are not involved.

Q: What are the most common viral causes of laryngitis?

A: Parainfluenza virus, influenza virus, and adenoviruses are the most common. Occasional causes are RSV, rhinoviruses, coronaviruses, and echoviruses. This is another one of those conditions that can have multiple etiologies (viruses, bacteria, environmental, allergic, chemical, etc.). These viruses, especially parainfluenza, are also common causes of laryngotracheobronchitis (barking cough).

Q: Where are the lesions of molluscum contagiosum typically found?

A: It is typically a cutaneous infection affecting the genital area of adults (it can be sexually transmitted); arms, legs, face, and trunk of children (it can be acquired by direct or indirect contact). The papules are commonly found in the axillae, the antecubital and popliteal fossae, and the crural folds. The genital lesions can be difficult to differentiate from other genital lesions (skin-colored papules with an opalescent character that are dome shaped, usually <10 mm).

Q: How is the virus that causes molluscum contagiosum classified?

A: Poxvirus. It is called the molluscum contagiosum virus.

Q: What causes pericarditis?

A: This is another long list. Viral causes include coxsackie viruses (probably number one cause) and echoviruses, as well as adenoviruses, mumps virus, and EBV. Endocarditis is most commonly a bacterial disease (especially *Staphylococcus aureus*, *Staphylococcus epidermidis*, and viridans streptococci).

Q: What do Orthomyxoviruses, Paramyxoviruses, and Rhabdoviruses have in common?

A: They are negative-stranded RNA viruses that carry their own RNA-dependent RNA polymerase within the virion.

Q: List some important Orthomyxoviruses, Paramyxoviruses, and Rhabdoviruses.

A: Orthomyxoviruses: influenza
Paramyxoviruses: measles, mumps, RSV, parainfluenza, human metapneumovirus
Rhabdoviruses: rabies, vesicular stomatitis virus (VSV; usually animals)

Q: A patient with AIDS died about five months after the slow onset of neurological symptoms (affecting speech, vision, coordination, etc.). Autopsy revealed areas of demyelination in the CNS. What is the most likely diagnosis and etiological agent?

A: The patient most likely died of progressive multifocal leukoencephalopathy due to polyomavirus infection. The JC virus (named after the patient from whom it was originally isolated) is the usual cause. This polyomavirus causes a latent infection in more than half of all adults. Severe immunosuppression can result in disease. Another polyomavirus, BK virus, is associated with a mild respiratory illness in children.

Q: A 9-month-old infant with respiratory disease (cough, labored breathing, and fever) was brought into the pediatrics clinic by her mother. What is the most likely etiological agent?

A: RSV. Aerosolized ribavirin is approved for treatment in severe cases for high risk patients.

Q: What viral disease has been eradicated from Earth?

A: Smallpox (variola virus). The World Health Assembly announced the complete eradication of smallpox in 1980. However, the Centers for Disease Control and Prevention in Atlanta and the Institute for Viral Preparations in Moscow have not destroyed their virus stocks. There are unconfirmed reports that some other countries might have the virus in their possession. We are getting close to eradicating polio.

Q: What causes cold sores?

A: Cold sores (herpetic stomatitis) are due primarily to HSV-1 but also to coxsackievirus. Stomatitis can have multiple causes in addition to viruses, including allergies, smoking, riboflavin deficiency, and *Candida* infection. Canker sores have an unknown etiology and are not related to cold sores.

Q: What causes hand-foot-and-mouth disease?

A: Usually coxsackievirus A16, but also some other enteroviruses. It is a common illness of infants and children, characterized by fever, sores in the mouth, and a rash with blisters. Do not confuse hand-foot-and-mouth disease with foot-and-mouth disease of hooved animals. The United States has been free of foot-and-mouth disease since 1929.

Q: How does yellow fever present?

A: Fever, chills, headache, vomiting (may be black due to hemorrhage of gastric mucosa), myalgia, backache, and proteinuria. Severe disease can present with jaundice, hemorrhagic manifestations, and renal failure.

Q: Yellow fever involves what organs?

A: Primarily the liver, but other organs (kidneys, spleen, lymph glands, and bone marrow) are also affected

Q: Where is yellow fever endemic?

A: Africa and South America

Q: What is the composition of the yellow fever vaccine?

A: Live attenuated virus

Q: Who should receive the yellow fever vaccine?

A: Persons at least 9 months of age who are traveling to or living in areas of South America and Africa where yellow fever infection is officially reported should be vaccinated. It is recommended for travel to countries that do not officially report the disease but that lie in the yellow fever–endemic zone.

Q: An electron microscopist had an electron micrograph of a bullet-shaped virus hanging on his wall. The virus was isolated from a patient who died from the infection. How do humans get infected with this virus?

A: Usually by an animal bite, most commonly a bat in the United States. It is most likely the rabies virus.

Q: Over the past year you have been checking one of your patients for serum HBsAg and have noted its continued presence. What can you conclude about this patient?

A: The patient is a chronic carrier of HBV. If HBsAg is detected and HBeAg is not, there will be minimum effect on the liver. If both HBsAg and HBeAg are detected, liver damage will continue.

Q: In general terms, what kind of infections does human bocavirus cause?

A: Respiratory, particularly children. It is classified in the family *Parvoviridae*. Infection is probably underreported.

Q: How is the West Nile virus classified?

A: It is a flavivirus, similar to the St. Louis encephalitis virus.

Q: What two nonhuman organisms are most important in the spread of West Nile fever?

    A: Mosquitoes (*Culex, Aedes,* and *Anopheles*) and birds

Q: How does West Nile fever present?

    A: Most infections are clinically inapparent or infection with sudden onset (~20%, high estimate) of mild fever (aka West Nile fever). Symptoms may include malaise, anorexia, nausea, vomiting, eye pain, headache, myalgia, rash, or lymphadenopathy. There are usually no significant CNS manifestations. The incubation period ranges from about 3 to 14 days with symptoms lasting about 3 to 6 days.

Q: What is the most significant risk factor for developing severe neurological disease after West Nile infection?

    A: Advanced age. Encephalitis is more common than meningitis. It can lead to a febrile illness with headache, mental confusion, tremors, or flaccid paralysis. As the disease continues, patients can progress to disorientation, decreased consciousness, stupor or coma, and death.

Mycology

Q: What are the two major groups of fungi?

A: Molds, which have mycelia, and yeasts, which are single celled

Q: What are dimorphic fungi? Give some examples.

A: Fungi that can exist as both a mold (room temperature) and a yeast (body temperature). Examples include *Histoplasma capsulatum*, *Blastomyces dermatitidis*, *Coccidioides immitis*, and *Paracoccidioides braziliensis*. *Sporothrix schenckii* and *Penicillium marneffei* are two other examples, but it is unlikely this characteristic for either one will be on most standardized tests. Anything about *P. marneffei* is unlikely to appear on standardized exams.

Q: What is the most common cause of fungal meningitis?

A: *Cryptococcus neoformans*. This is usually called cryptococcal meningitis.

Q: What laboratory test is commonly used to identify *Candida albicans*?

A: The germ tube test. This yeast forms germ tubes (long extensions that resemble nonseptate hyphae) after incubation for several hours in serum. Germ tubes should not be confused with pseudohyphae. Pseudohyphae can form in culture and during infection and look like yeast cells with a chain of budding cells.

Q: Differentiate between superficial, cutaneous, subcutaneous, and systemic mycoses.

A: Superficial mycoses affect the outermost layers of skin and hair. There is usually no stimulation of the immune system. Cutaneous mycoses affect the keratinized layers of the skin, hair, and nails at a deeper level than the superficial mycoses. Stimulation of the immune system is common. Subcutaneous mycoses obviously affect deeper levels of the dermis and can involve muscle or bone. The systemic mycoses include some of the more serious mycotic infections and can involve many of the internal organs. These fungi can cause disease in immunocompetent as well as immunosuppressed individuals.

Q: Where in the United States is coccidioidomycosis most prevalent?

A: Southwestern United States (south central California, Nevada, Arizona, New Mexico, and western Texas). It is treated with amphotericin B, itraconazole, or fluconazole. Note: Amphotericin B is always a good guess for therapy of fungal infections. An alternative drug may be a better choice if nephrotoxicity is an issue.

Q: What is the mode of action of amphotericin B?

A: This drug also targets ergosterol, but its mode of action is different from the azoles. Amphotericin B is a polyene that acts by binding to ergosterol in the fungal cell membrane, resulting in cell leakage and death.

Q: Name some superficial mycoses (hair and outer layers [stratum corneum] of skin) and their etiological agents.

A: The only one you most likely need to know is pityriasis versicolor (formerly tinea versicolor; skin infection) caused by *Malassezia furfur* (synonym *Pityrosporum orbiculare*). Others include tinea nigra (skin; *Hortaea werneckii*, formerly known as *Phaeoannellomyces werneckii*, formerly known as *Exophiala werneckii* and *Cladosporium werneckii*), Black piedra (hair; *Piedraia hortae*), and White piedra (hair; *Trichosporon asahii* [formerly *Trichosporon beigelii*]). For these last three, just recognize them as superficial mycoses. You will not need to know their etiological agents. Note: There are at least 10 species of *Malassezia*, but *M. furfur* is most common in the literature.

Q: How are superficial mycoses definitively diagnosed?

A: Microscopically. Skin scrapings are first treated with an alkali such as potassium hydroxide that dissolves the skin, making it easier to see the fungi. *Malassezia furfur* has a characteristic "spaghetti and meatball" appearance (clusters of hyphae and yeast cells; it is dimorphic and you can see both forms). *M. furfur* can cause hyper- or hypopigmentation of the skin, especially the shoulders and upper trunk. The organism is lipophilic and most commonly causes infections in teenagers and young adults.

Q: What is the treatment for superficial mycoses?

A: Skin infections can be treated with a number of different topicals, including miconazole. Infected hair can be cut off.

Q: Name some cutaneous mycoses (keratinized layers) and their etiological agents.

A: These infections affect the skin, nails, and hair but at a deeper level than the superficial mycoses, although these are also commonly referred to as superficial fungal infections or superficial tinea infections. A fungal infection of the skin is also called a dermatophytosis, tinea, or ringworm. The primary fungi are various species of *Trichophyton, Microsporum,* and *Epidermophyton.* Infections caused by these organisms are commonly named for the sites involved: tinea capitis (dermatophyte infection of the head), tinea barbae (beard area), tinea corporis (body surface), tinea manuum (hands), tinea pedis (feet), and tinea unguium (toenails). These names do not distinguish between species (e.g., tinea capitis may be caused by *Trichophyton* or *Microsporum* species). Because they can grow on keratin, these fungi are referred to as being keratinophilic.

Q: How are cutaneous mycoses diagnosed?

A: Direct microscopic observation of skin scrapings, similar to the superficial mycoses using 10% potassium hydroxide as a clearing agent. A Wood's lamp (ultraviolet light) is typically not too useful for skin infections. Identification of the specific organism requires growth on laboratory media, such as Sabouraud dextrose agar, but this is slow and rarely required.

Q: What is the treatment for cutaneous mycoses?

A: There are a number of topicals, usually creams, lotions, or powders, containing one of several active antifungals (miconazole, co-trimoxazole, tolnaftate, etc.).

Q: What fungi are involved in dandruff production?

A: *Malassezia* species. Sebaceous gland secretions, particularly fatty acids, serve as a nutritional source for *Malassezia.* It can be treated with shampoos containing selenium or ketoconazole. The individual response to these agents is widely variable.

Q: Name some subcutaneous mycoses and their etiological agents.

A: The most common in the United States is sporotrichosis, due to *Sporothrix schenckii.* There are others such as chromoblastomycoses, phaeohyphomycosis, and eumycotic mycetoma, but it is unlikely you need to know any of them or their causes for any standardized exam.

Q: What is the route of infection for subcutaneous mycoses?

A: Usually some kind of trauma, such as a thorn or splinter. For *Sporothrix schenckii,* it is commonly associated with a puncture from a rose thorn. It grows on the thorns, but it can be found elsewhere in the environment. Infections have been associated with contaminated soil, cats or some wild animals, or by inhalation of spores. The disease is characterized by nodular lesions of cutaneous and subcutaneous tissues and adjacent lymphatics that suppurate and ulcerate.

Q: How are subcutaneous mycoses diagnosed?

A: It depends on the etiological agent. Specimens, including tissue biopsies or draining purulent fluid, should be sent to the lab for histological examination and culture. Some bacteria, such as the anaerobe *Actinomyces,* can mimic subcutaneous mycoses.

Q: What is the treatment for subcutaneous mycoses, such as sporotrichosis?

A: Sporotrichosis can be treated with itraconazole or oral potassium iodide. Amphotericin B is sometimes used for serious disease (e.g., central nervous system involvement). For some subcutaneous mycoses, surgery may be required.

Q: What are some azole drugs that are effective against fungi?

A: Fluconazole, itraconazole, voriconazole, and ketoconazole

Q: What sterol, found in the cytoplasmic membranes of fungi, is the target for azole drugs?

A: Ergosterol. The azoles block the activity of 14-alpha-demethylase, leading to inhibition of ergosterol synthesis. Mammals have cholesterol in their membranes, so human cells are not affected.

Q: Name some systemic mycoses of immunocompetent individuals and their etiological agents.

A: Histoplasmosis (*Histoplasma capsulatum*), blastomycosis (*Blastomyces dermatitidis*), paracoccidioidomycosis (*Paracoccidioides brasiliensis*), coccidioidomycosis (*Coccidioides immitis*), and cryptococcosis (*Cryptococcus neoformans*). As you might expect, these same diseases are more common in immunosuppressed individuals.

Q: From the previous question, what is the geographical distribution of each disease?

A: It is possible to become infected with each of the organisms in many different geographical areas, but some endemic areas have a higher incidence. The following is a list of some of the endemic areas:

- Histoplasmosis: Mississippi and Ohio River valleys in the midwest United States (learn this)
- Blastomycosis: primarily North America (widely distributed, including the Mississippi and Ohio River valleys) and parts of Africa
- Paracoccidioidomycosis: Central and South America, especially Brazil
- Coccidioidomycosis (valley fever): Lower Sonoran life zones (arid portions of southwest United States [California to Texas; most cases in the United States are in the San Joaquin Valley of California and south central Arizona], including parts of Mexico and South America)
- Cryptococcosis: worldwide

Make a table of these to help you learn them.

Q: Cryptococcosis is a zoonotic disease. What animal is the carrier and how do humans get infected?

A: Humans are infected by inhalation of dust containing dried pigeon and other bird droppings.

Q: What is the most common clinical presentation of cryptococcosis?

A: Although *Cryptococcus neoformans* can cause pulmonary infections, it is usually asymptomatic at this stage. Some patients may develop a severe pneumonia. The most common clinical presentation is meningitis after hematogenous spread from the lungs, but cutaneous involvement is also common. Disseminated disease, especially to the central nervous system, is common in AIDS patients.

Q: How is cryptococcal meningitis diagnosed?

A: There are several serological tests available (latex agglutination and enzyme-linked immunosorbent assay), but the most rapid test is the microscopic India ink test. *Cryptococcus neoformans* is a yeast with a polysaccharide capsule. Because the capsule excludes the India ink particles, the yeast cells in cerebrospinal fluid can be seen with a "halo." Confirmation, if required, can be done by culture on Sabouraud dextrose agar.

Q: From the list of systemic mycoses, what other disease has been associated with birds?

A: Histoplasmosis, kind of. This is a trick question. Birds do not get infected and do not transport the organism. Bird droppings (and bat guano) serve to enrich the soil so that the mycelial form of the organism can grow in the soil. As with cryptococcosis, it is also acquired by inhalation of contaminated dust. Bats are carriers of the organism. Therefore if a question pops up on a standardized exam concerning bats (e.g., someone who becomes ill after exploring a cave), you need to consider histoplasmosis as well as rabies. *Histoplasma capsulatum* is a dimorphic fungus that grows as yeast in the body (as do other dimorphic fungi).

Q: How does histoplasmosis present?

A: Asymptomatic in most (>99%), but mild flu-like symptoms can occur. Various other pulmonary syndromes (e.g., chronic cavitary, granulomatis mediastinitis, and others) have been described. Systemic disease, usually in the immunosuppressed, can be life threatening.

Q: How does one get infected with *Coccidioides immitis*?

A: Inhalation of dust containing the organism. As stated previously, the endemic area in the United States is the southwestern arid region, especially the San Joaquin Valley in California and south-central Arizona (learn this).

Q: How does coccidioidomycosis present?

A: About 40% of infected patients are symptomatic, usually with mild pulmonary symptoms resembling self-limiting community acquired pneumonia. Serious systemic disease can occur with an incidence of less than 5% of symptomatic patients.

Q: How is coccidioidomycosis diagnosed?

A: In tissues, it forms endospore-containing spherules that can be seen microscopically. Visualization of these is diagnostic. Potassium hydroxide wet mounts are useful but lack the sensitivity and specificity of several other staining methods, such as the calcofluor white fluorescent stain. A number of serological tests are available for detection of *Coccidioides*-specific IgG and IgM in patients' sera. Eosinophilia is common.

Q: What are the clinical manifestations of blastomycosis?

    A: Usually asymptomatic, but pulmonary symptoms can occur, presenting as acute or chronic pneumonia. Extrapulmonary disease can involve any organ, but cutaneous lesions are most common.

Q: Describe the infectious form of *Blastomyces dermatitidis* as it would look microscopically.

    A: Yeast with broad-based budding cells. This is another dimorphic fungus that grows as a mold (hyphae) in the environment.

Q: Name some systemic mycoses of immunocompromised individuals and their etiological agents.

    A: Primarily candidiasis (*Candida albicans* and several other *Candida* species) and aspergillosis (*Aspergillus* spp.) are the most important, but you should be aware of zygomycosis (mucormycosis; *Rhizopus* spp. and several others), and fusariosis (*Fusarium* spp.).

Q: Who is at risk for opportunistic fungal infections?

    A: As with other opportunistic pathogens, the fungi that cause opportunistic infections do so in immunocompromised individuals. These mycoses can be superficial or systemic. The infections commonly referred to as the systemic mycoses are usually caused by fungi that cause infections in immunocompetent individuals (they are true pathogens with high virulence).

Q: Name some opportunistic fungal pathogens.

    A: This list includes those listed above for the systemic mycoses of immunocompromised individuals including *Candida albicans* and other species of *Candida* (candidiasis), *Aspergillus* (aspergillosis), *Rhizopus* and several other genera (zygomycosis), *Fusarium* spp. (fusariosis), and *Pneumocystis jiroveci* (*carinii*).

Q: What is the environmental source of *Candida albicans*?

    A: It is part of our normal flora. It can be found in the oral cavity, intestinal tract, and vagina.

Q: What test is used to differentiate *C. albicans* from other species of *Candida*?

    A: Germ tube test

Q: *C. albicans* infections of the skin are most commonly located where?

    A: In the intertriginous areas (i.e., in the folds of the skin, axilla, groin, under the breasts, etc.). *C. albicans* infection appears as a beefy red rash with satellite lesions.

Q: What fungi are communicable?

    A: Not many. *Candida* and the dermatophytes (*Trichophyton*, *Microsporum*, and *Epidermophyton*) are about the only ones.

Q: What is the most common opportunistic infection associated with AIDS?

    A: *Pneumocystis jiroveci* (*carinii*) pneumonia. *P. jiroveci* has been difficult to classify. It was originally classified as a parasite, but it is actually a fungus. The name of the human pathogen, *Pneumocystis carinii*, was changed to *Pneumocystis jiroveci* in 2001. *Pneumocystis carinii* still exists, but it is found only in rats. Other common opportunistic pathogens in AIDS include the *Mycobacterium avium* complex, *Toxoplasma* (encephalitis), cytomegalovirus, *Cryptococcus neoformans*, and *Candida* spp.

Parasitology

Q: In parasitology, what is meant by the intermediate host and the definitive host?

A: The intermediate host is the organism that houses the immature (nonsexually reproducing) stage of a parasite. The definitive host is the organism that houses the mature (sexually reproducing) stage.

Q: What are helminths?

A: Worms

Q: What are the three major classes of helminths?

A: Nematodes, trematodes, and cestodes

Q: Describe the nematodes and give examples.

A: These are the roundworms, such as *Enterobius vermicularis, Ascaris lumbricoides, Trichuris trichiura, Necator americanus, Strongyloides stercoralis,* and *Trichinella spiralis.*

Q: What is the most common helminth infection in the United States?

A: *Enterobius vermicularis*

Q: What infection does *Enterobius vermicularis* cause?

A: Pinworm infection (enterobiasis)

Q: What is the most common symptom of pinworm infection?

A: Perianal itching. Initial infection is by ingestion of the eggs, but autoinfection can also occur when the eggs hatch and the worms migrate into the rectum.

Q: How is *Enterobius* infection diagnosed?

A: The Scotch tape test. The sticky side is applied to the perianal area followed by smoothing the tape out on a glass slide using a cotton swab. The slide is examined for eggs (ova) with the microscope set on low power.

Q: What do the pinworm ova look like?

A: They are relatively large, oval, and have a thick outer shell. One side is flattened relative to the other.

Q: Infection with what tapeworm can result in vitamin B_{12} deficiency?

A: *Diphyllobothrium latum*, the fish tapeworm. This longest of all tapeworms (more than 20 feet long) can infect humans (raw, undercooked, or pickled fish) and may compete with the host for vitamin B_{12}. Most infections are asymptomatic, but megaloblastic anemia can result (very rare).

Q: The Duffy blood group antigens serve as receptors for infection due to what parasite?

A: *Plasmodium vivax*. People who do not have these antigens (most West Africans) are not at risk for infection from this species of malaria parasite but can still get malaria from other species. Those with the Duffy coat antigens (most of European lineage) are at risk.

Q: Cat litter boxes should *not* be changed by whom and why?

A: Pregnant women. Cats are the definitive hosts for *Toxoplasma gondii*, the parasite that causes toxoplasmosis. *T. gondii* can cross the placenta and infect the fetus, possibly resulting in miscarriage, stillbirth, or severe neonatal complications such as blindness and mental retardation. Most congenital infections are asymptomatic. The fetus is at risk only if the mother gets infected during pregnancy. The number of infants born with toxoplasmosis in the United States is not known but has been estimated to be between 400 and 4000. The parasite typically has no effect on healthy adults. In fact, the Centers for Disease Control and Prevention estimates that 60 million people in the United States are infected. Immunosuppressed individuals, including those with AIDS, can have severe life-threatening infections.

Q: What is the difference between a trophozoite and a cyst?

A: The trophozoite is the motile feeding form of protozoan parasites, and the cyst is the inactive resistant form (analogous to bacterial spores). Both forms can be infectious. A cyst form has not been found in some species (e.g., *Trichomonas vaginalis*). For protozoans that infect by ingestion, assume the cyst is the infectious form and you will usually be right.

Q: List some important parasitic protozoans.

A: *Entamoeba histolytica, Giardia duodenalis* (syn. *G. lamblia* or *G. intestinalis*), *Trichomonas vaginalis, Cryptosporidium*, and *Cyclospora* are some, but there are others.

Q: How is *Ascaris lumbricoides* classified, what disease does it cause, and what is the major complication?

A: It is a roundworm that causes ascariasis. Ascariasis is not common in North America. It can lead to intestinal blockage by mature adult worms.

Q: What is the infectious form of *Ascaris lumbricoides*?

A: Egg. The eggs of *A. lumbricoides* are distinctive looking, having an oval mammillated (small rounded protuberances on their surface) appearance.

Q: Briefly describe the life cycle of *A. lumbricoides*.

A: Eggs from contaminated soil are ingested; larval worms are released and penetrate duodenal wall; hematogenous spread to heart; entrance into pulmonary circulation leads to infection of alveoli of lungs; ascend trachea by coughing; subsequent swallowing returns larvae to small intestine where worms mature into adults. Eggs are passed in feces.

Q: What is the treatment for *A. lumbricoides* infection?

A: Albendazole (drug of choice) or alternatively mebendazole.

Q: What is the common name for *Necator americanus* and *Ancylostoma duodenale*?

A: Hookworm. *N. americanus* is the New World hookworm, and *A. duodenale* is the Old World hookworm. There is no significant difference in infections caused by the two species. They are difficult to speciate, and their eggs are virtually identical microscopically.

Q: Briefly describe the life cycle of hookworms.

A: It is similar to that described above for *Ascaris lumbricoides*. The major difference is that infection is by larvae (filariform) found in warm soil. The infectious larvae can penetrate intact skin, after which they enter the circulation and eventually reach the lungs. After migration via the trachea, the larvae are swallowed and the worms mature in the small intestine. The eggs are passed with feces into the soil where they hatch, commonly within 24 hours, and develop (within another 24 hours) into the infective filariform (larvae). Iron deficiency anemia is a common feature of hookworm disease.

Q: How do most people get infected with hookworms?

A: Walking barefoot on infected soil. Infection can lead to anemia due to iron deficiency.

Q: What is the drug of choice for hookworm infection?

A: Albendazole or mebendazole are the two drugs of choice, either alone or with praziquantel.

Q: How is cryptosporidiosis acquired?

A: Fecal–oral, commonly by drinking contaminated water. The infectious cysts (oocytes) are very hardy and difficult to kill, even with chlorine. Contaminated public swimming pools require increased time and concentration of soluble chlorine for disinfection.

Q: How do initial infections with *Strongyloides stercoralis* occur?

A: *S. stercoralis* is similar to the hookworms in that filariform larvae in the soil can penetrate intact skin.

Q: Briefly describe the infectious cycle of *Strongyloides*.

A: The infectious cycle is similar to that described above for hookworms. An important difference is that the eggs laid by *Strongyloides* can hatch in the intestinal tract (rather than being passed with feces to the soil as with hookworms and *Ascaris lumbricoides*) and develop into rhabditiform (noninfectious) larvae. The rhabditiform larvae can then either be passed in the feces (and later develop into infectious filariform larvae) or they can develop into infectious filariform larvae in the intestine where they can cause a reinfection (autoinfection). The continuation of the autoinfection cycle can lead to hyperinfection, especially in the immunocompromised, and can have a high mortality rate. Infection in most people is commonly asymptomatic, but the infection can also cause mild gastrointestinal symptoms. In the United States it is most common in Kentucky, Tennessee, and West Virginia.

Q: What does *Trichuris trichiura* cause?

A: Whipworm infection, usually in children. The disease is usually asymptomatic. One of the more serious manifestations is rectal prolapse. It is spread by the fecal–oral route.

Q: How is the diagnosis of whipworm infection made?

A: Diagnosis is based on the identification of the distinctive eggs in stools. In the United States it is most common in the southeastern states.

Q: Describe the morphology of whipworm eggs.

A: Eggs of *Trichuris trichiura* have a symmetrical elongated shape with distinctive polar plugs at each end. You should look for an image of these eggs. Once you see them, they are easy to recognize.

Q: List some parasites for which metronidazole is the drug of choice.

A: *Giardia duodenalis* (syn. *G. lamblia* or *G. intestinalis*), *Trichomonas vaginalis*, and *Entamoeba histolytica* are the three that you should know. Giardiasis can also be treated with tinidazole.

Q: An outbreak of diarrhea at a child daycare center was traced to a "water table" where most of the children played each day. One of the children was taken to his doctor. An O & P (ova and parasites) exam performed on a stool specimen revealed small, flagellated, pear-shaped cells that appeared to have two "eyes." What is the most probable diagnosis?

A: Giardiasis. The important epidemiological information is that this occurred at a daycare center. The other major clue is the description of the organism, which is classic for the trophozoite of *Giardia duodenalis*. The two "eyes" are nuclei. It is important that you look at some images of this organism if you do not believe you can recognize it microscopically. There is a good possibility that an image will be on your standardized exams.

Q: From the previous question, what would your diagnosis be if the cells seen microscopically were more rounded, had no flagella or "eyes," and their color was red by the modified acid-fast stain?

A: Your best guess would be *Cryptosporidium parvum*. This protozoan, along with *Giardia duodenalis*, is associated with outbreaks of diarrhea in daycare facilities. Make sure you know that *Cryptosporidium* is acid-fast and that acid-fast organisms (parasites and bacteria) appear red microscopically. If you really know your parasitology, you might also guess *Cyclospora*, but not much is known about its epidemiology. Most standardized exams have no questions on *Cyclospora*.

Q: What parasite has been most commonly associated with eating sushi (raw fish)?

A: The best guess with no further information is the fish tapeworm, *Diphyllobothrium latum*. Raw or undercooked fish can be the source of a number of other parasites, including *Clonorchis sinensis* (Chinese or oriental liver fluke, now called *Opisthorchis sinensis*). Choose the fish tapeworm if the infection originates in North America or the oriental liver fluke if the infection originates in Asia (such as an Asian immigrant).

Q: What are the symptoms of giardiasis?

A: Diarrhea, abdominal cramps, and nausea are the most common symptoms. Symptoms, which develop 1–2 weeks after ingestion of the parasite, can be mild to severe. Diarrhea can have an acute onset and is often foul smelling, frequently high in fat (steatorrhea). After infection, it inhabits the small bowel and adheres to the epithelium, but it does not invade tissue. Although the exact mechanism of pathogenicity is not known, it does not appear to secrete an enterotoxin.

Q: Describe the epidemiology of giardiasis.

A: The major points are that wild animals, such as beavers and muskrats, are reservoirs and that fecal contamination of freshwater lakes and streams can lead to infection in humans if the water is consumed. After initial infection, the fecal–oral route can lead to other people becoming infected. This organism produces cysts that can resist drying in the environment and that are the infective forms. Questions about *Giardia* infections often use camping or backpacking in a mountainous region as a theme.

Q: Describe the trematodes and give examples.

A: These are the flukes or flat worms. Examples include the blood flukes (*Schistosoma mansoni* [most important to know], *S. japonicum*, and *S. haematobium*), *Clonorchis* (*Opisthorchis*) *sinensis* (Chinese liver fluke), and *Paragonimus westermani* (lung fluke).

Q: Briefly describe the life cycle of flukes.

A: You should know that all flukes pass through an intermediate host (snail is the keyword here) and may or may not have a secondary host.

Q: Which species of *Schistosoma* is found in the Western Hemisphere, including Puerto Rico?

A: *Schistosoma mansoni*

Q: How is a person infected with schistosomes?

A: Snails release infectious cercariae (larval form) that can penetrate intact skin. The cercariae penetrate the skin and travel through the heart, lungs, and circulatory system until they reach the portal veins where they develop into nonpathogenic worms. The infectious form of all other flukes is ingested from infected fish or crustaceans or on vegetation.

Q: How is schistosomiasis diagnosed?

A: By microscopic detection of the characteristic eggs (ova) in stools. The eggs are released by the female worm.

Q: Describe the eggs of *Schistosoma*.

A: The large oval eggs have a prominent lateral spine, especially visible with *S. mansoni*. This is another image you should review.

Q: Why is host resistance in humans to *Schistosoma* so poor?

A: This parasite coats itself with substances that the human host recognizes as self.

Q: Ingestion of plants that grow in water (e.g., watercress or water chestnuts) is associated with infections due to what parasites?

A: Flukes such as *Fasciola hepatica* (watercress) and *Fasciolopsis* (water chestnuts). These parasites are rare in the United States.

Q: Ingestion of freshwater crabs or crayfish (crustaceans) is associated with infection due to what trematode?

A: *Paragonimus westermani*, the lung fluke. The parasite is rare in the United States.

Q: What is dracunculiasis?

A: Dracunculiasis, also called guinea-worm disease, is caused by a worm, *Dracunculus medinensis,* and is found primarily in rural Africa. *D. medinensis* is the largest of the tissue parasites affecting humans. During an infection the worm migrates through subcutaneous tissues, causing severe pain, especially in the joints. The worm eventually emerges, usually from the feet, causing an intensely painful edema, a blister, and an ulcer accompanied by fever, nausea, and vomiting. The World Health Organization has an initiative to eliminate dracunculiasis from Earth.

Q: Describe the cestodes and give examples.

A: The cestodes are the tapeworms. They consist of a head (scolex) to which a chain of flat segments called proglottids are attached. Examples include *Taenia saginata* (beef tapeworm), *Taenia solium* (pork tapeworm), and *Diphyllobothrium latum* (fish tapeworm). For these tapeworms, humans become infected after eating raw or undercooked beef, pork, or fish. Note: Rare steak or medium-cooked pork are generally safe to eat in the U.S.

Q: What is the function of the scolex and the proglottids?

A: The scolex attaches to the intestinal wall. The proglottids are the reproductive segments. Tapeworms are hermaphroditic, with both male and female reproductive structures in each proglottid. Gravid segments containing eggs are shed in the stools.

Q: What is cysticercosis?

A: Human infection with the pork tapeworm, *Taenia solium*. It is usually due to ingestion of undercooked food in which the infectious larvae (cysticerci) in muscle are not killed. After ingestion, the cysticercus attaches to the intestinal wall and grows as a typical tapeworm. In cysticercosis, infection is due to ingestion of tapeworm eggs, such as from contaminated vegetables or water. In this case, after the eggs hatch, the free embryo penetrates the intestinal wall where it becomes blood-borne. Although usually asymptomatic, infection of various tissues, including the brain (neurocysticercosis), can occur with serious consequences.

Q: What is the vector for malaria?

A: *Anopheles* mosquito

Q: Name the four species of malaria parasite.

A: *Plasmodium vivax, P. ovale, P. malariae,* and *P. falciparum*

Q: Which malaria parasite has the widest geographical distribution?

A: *Plasmodium vivax.* It is found in temperate, subtropical, and tropical areas.

Q: Which malaria parasite causes the most serious form of disease?

A: *Plasmodium falciparum.* This organism has the shortest incubation period (about one week) and the highest mortality rate than the other species.

Q: Briefly describe the life cycle of *Plasmodium* that results in malaria.

A: Sporozoites (infective forms) in mosquito saliva are injected through skin and enter the circulation. In the liver, the sporozoites invade hepatocytes and differentiate into merozoites. The merozoites are released into the circulation where they enter erythrocytes (initiating the erythrocytic cycle). (In the liver, *P. vivax* and *P. ovale* can develop into a latent form [called the hypnozoite] that is responsible for relapses). In the erythrocytes, the parasite goes through various stages of the asexual cycle (called schizogony) from the ring stage, to trophozoite, to the merozoite-filled schizont. Lysis of erythrocytes releases merozoites into the circulation. The cycle repeats with reinfection of erythrocytes by merozoites. Release of merozoites corresponds to the recurrent symptoms seen every two days for *Plasmodium vivax* and *P. falciparum.* In some erythrocytes, a sexual cycle produces male and female gametocytes. These erythrocytes can be ingested by mosquitoes in which a sexual cycle produces the infectious sporozoites.

Q: Describe the general symptoms of malaria.

A: After the incubation period (about one week for *P. falciparum* and two weeks for *P. vivax*), the disease eventually enters the two-day cycle (tertian malaria) of flu-like symptoms consisting of fever and chills. Splenomegaly and anemia are common. *P. falciparum* can have more severe manifestations (see next question).

Q: What are the severe manifestations of *P. falciparum* infection?

A: Anemia, capillary blockage, kidney failure, cerebral malaria, and death

Q: What is blackwater fever?

A: *P. falciparum* infection in which kidney damage and the severe anemia due to erythrocyte destruction cause the urine to turn a dark color

Q: How is the laboratory diagnosis of malaria made?

A: Microscopic evaluation of Giemsa-stained blood smears, looking for ring forms (trophozoites within red blood cells), gametocytes, and free parasites

Q: What is the treatment for malaria?

A: Chloroquine is the drug of choice, but resistance is developing. Primaquine can be used to prevent relapses due to *P. vivax.* Quinine, mefloquine, pyrimethamine-sulfadoxine, or several other drugs can be used for chloroquine-resistant strains.

Q: What is the vector for sleeping sickness?

A: Tsetse fly. Also called African trypanosomiasis, the disease is fatal if untreated.

Q: What parasite causes sleeping sickness?

A: *Trypanosoma* is probably close enough. The taxonomy of these trypanosomes gets confusing. There is one species of *Trypanosoma brucei* and three subspecies: *Trypanosoma brucei gambiense, Trypanosoma brucei rhodesiense,* and *Trypanosoma brucei cruzi.* These are often written without the middle "*brucei.*" The first two of these cause sleeping sickness.

Q: How are the trypanosomes classified?

A: They are protozoa and along with *Leishmania* are referred to as the hemoflagellates.

Q: What does *Trypanosoma brucei cruzi* cause?

A: Chagas (pronounced sha'gus) disease. You will usually see this organism written as *Trypanosoma cruzi.*

Q: What is the vector for Chagas disease?

A: Species of reduviid bugs. Within this large family are species such as *Triatoma* and *Paratriatoma.* These are sometimes called kissing bugs because they often bite on the face, including near the mouth.

Q: What is another name for Chagas disease?

A: American trypanosomiasis. It is most common in Mexico and Central and South America. It is rare in the United States.

Q: Exposure to cat feces can result in what serious parasitic infection in pregnant women?

A: Toxoplasmosis, which can result in premature delivery, permanent neurological sequelae in newborns, spontaneous abortion, or stillbirth. Most cases of toxoplasmosis in men, women, and children are asymptomatic. Diagnosis is usually by serological testing.

Q: What is meant by an intermediary host?

A: In parasitology, an intermediate host is one in which the parasite multiplies asexually. The sexual stage is in the definitive host.

Q: What is the vector of malaria?

A: *Anopheles* mosquito

Q: What is the vector of onchocerciasis (*Onchocerca volvulus*)?

A: Blackfly

Q: Where is onchocerciasis endemic?

A: Africa, South America, and the Middle East

Q: What are the major symptoms of onchocerciasis?

A: Development of subcutaneous nodules. The worms (microfilariae) can migrate to the eyes and cause blindness due to scarring of the cornea. The disease is also called river blindness.

Q: Sandflies are vectors for what disease?

A: Leishmaniasis

Q: Where is leishmaniasis endemic?

A: It depends on the species of *Leishmania*. There are three species: *L. donovani* (most important to know about; Asia and Africa), *L. tropica* (Asia, Africa, southern portions of Europe, and the former Soviet Union), and *L. braziliensis* (Central and South America). You do not need to know these names to the species level, just the genus *Leishmania*.

Q: How does leishmaniasis present?

A: It can be subclinical (inapparent), localized (skin lesions), or disseminated (cutaneous, mucosal, or visceral). Cutaneous and visceral are the most common forms of active disease. Cutaneous disease (*L. tropica* and *L. braziliensis*) begins as macules or papules, developing into erythematous skin ulcers. The ulcers heal very slowly over a matter of months. Visceral leishmaniasis (primarily do to *L. donovani*) presents with fever, weight loss, enlargement of the spleen and liver, and anemia. If left untreated, it is nearly always fatal.

Q: What protozoan disease is tick-borne in the United States?

A: Babesiosis, a rare disease spread by the bite of *Ixodes scapularis,* the same tick that is the vector for Lyme disease and human granulocytic anaplasmosis. *Babesia microti* is the most common bacterial species responsible for babesiosis. As with malaria, the parasite infects erythrocytes. Most reported cases occur during the summer months along the coast and off-shore islands of the Northeast.

Q: What causes scabies?

A: Mites (*Sarcoptes scabiei*). The mites burrow into the skin (not below the stratum corneum) and result in itching due to a delayed hypersensitivity reaction. It causes mange in animals.

Q: What is *Hymenolepis nana*?

A: It is the tapeworm that most commonly infects humans in the United States, often due to contamination of grains, including flour. The tapeworm is autoinfective and can be spread via the fecal–oral route. It can be spread to foods via some insects (beetles) and mice.

Q: What is the intermediate host for *Hymenolepis nana*?

A: None. This is the only human tapeworm that has no intermediate host.

Q: What is the cause of hydatid disease?

A: Hydatid disease (echinococcosis) is caused by the tapeworm *Echinococcus granulosus* or *E. multilocularis*. The life cycle includes both canines (definitive host: foxes, dogs, coyotes, etc.) and herbivores (intermediate host), particularly sheep. Humans become infected after consumption of food or water that is contaminated with feces. The disease is rare in the United States.

Q: Describe the pathophysiology of hydatid disease in humans.

A: The hydatid cyst usually develops in the liver or lungs of an intermediate host (the human in this case). As the cyst grows, it can exert pressure on adjacent tissues and organs, resulting in necrosis. Most patients remain asymptomatic. The cyst occasionally breaks, releasing a fluid that can cause anaphylactic shock. In many infected patients, years may pass before any symptoms become apparent. Alveolar echinococcosis is due to infection with the larval stage of *Echinococcus multilocularis*.

Q: What is the most common parasitic infection of humans in North America?

A: Pinworm, with 20–40 million cases in the United States. Other common parasitic diseases in the United States are giardiasis, trichomoniasis, and toxoplasmosis (although infection with *Toxoplasma gondii* is usually asymptomatic).

Q: How do people get infected with pinworm?

A: It is spread by the fecal–oral route. The perianal itching and subsequent scratching, especially among children, facilitates its spread. Activities related to its dissemination include nail biting, thumb sucking, poor hygiene, and inadequate hand washing.

Q: How is pinworm (*Enterobius vermicularis*) infection treated?

A: Typically when diagnosed, all infected family, class members, and other close contacts are treated. Antihelmintic drugs, such as albendazole or mebendazole (similar as for ascariasis), are effective. Personal hygiene improvements are required to end the infectious cycle.

Q: What is the common name for *Phthirus pubis*?

A: Pubic or crab louse. In all lice species, the females attach their ova (nits) to hair shafts or clothing.

Q: What is the agent of head lice?

A: *Pediculus humanus capitis*. The small white nits can be seen with the naked eye. Diagnosis of pediculosis is made by observation of the nits on hair shafts.

Q: What is the most common *worldwide* cause of hemolytic anemia?

A: Malaria

Q: How does a woman with trichomoniasis present?

A: She has a recent history of vaginal itching and burning, a thin yellowish-green bubbly discharge, and petechiae on the cervix ("strawberry cervix").

Q: In the laboratory, how is trichomoniasis diagnosed?

A: Microscopically, looking for the motile, pear-shaped protozoan (*Trichomonas vaginalis*) with its characteristic four flagella (It is a flagellate.)

Q: What is the drug of choice for *Trichomonas vaginalis*?

A: Metronidazole

Q: Can males have trichomoniasis?

A: Yes, but they are usually asymptomatic carriers. It can present as a urethritis with discharge.

Q: What is the cause of amebiasis and how does it present?

A: *Entamoeba histolytica* most commonly causes a bloody diarrhea, usually presenting as amebic colitis. There are other accompanying symptoms that you would see in any kind of colitis (cramps, fever, etc.). In the United States the disease is most common in immigrants. Liver abscess is a rare severe manifestation. Other than malaria, it is the second most common cause of protozoal death in the world.

Q: What infection with a high mortality rate is caused by *Naegleria fowleri*?

A: Meningoencephalitis. Infection can occur while swimming in fresh water. The infection rate is very low.

Q: What amoeba has been associated with keratitis in contact lens users?

A: *Acanthamoeba*. In severe cases, corneal transplantation may be required. In the United States 85% of cases occur in contact lens users. *Acanthamoeba* keratitis outbreaks have been associated with contaminated lens solution. The organism is ubiquitous in nature.

Q: What is the cause of visceral larval migrans?

A: *Toxocara canis* and *T. catis,* roundworms of dogs and cats, respectively. The disease is also called toxocariasis and has two main clinical presentations, visceral larva migrans and ocular larva migrans. As the term "migrans" suggests, these roundworms migrate or travel throughout the body, causing infections in various organs. Humans get infected by ingestion of contaminated soil.

Q: How are humans most frequently infected with *Trichinella spiralis*?

A: Historically in the United States, ingestion of improperly cooked pork; currently, ingestion of raw or undercooked wild animals

Q: Where in the body can the encysted larva of *T. spiralis* be found?

A: Striated muscle. Calcification may occur.

Q: What is *Loa loa*?

A: It is a nematode spread by a biting fly (*Chrysops*) in parts of equatorial West Africa. Infection can affect the eye, with the worms sometimes migrating across the eye. The worms can also cause subcutaneous swellings.

Q: What causes elephantiasis?

A: Two different parasites: *Wucheria bancrofti* and *Brugia* spp. The infection is also called lymphatic filariasis. Obstruction leads to lymph node enlargement. The microfilariae are highest in the circulation at night.

Q: How are *Wucheria* and *Brugia* infections acquired?

A: The larvae of these two nematodes are spread by mosquitoes. It is most common in tropical regions worldwide.

Q: What is kala azar?

A: Another name for visceral leishmaniasis. It is caused by *Leishmania donovani*. Behind malaria, it is the second most common cause of death due to parasites. It is also called black fever.

Random Pearls

Q: What is the most common bacterium associated with infective endocarditis in intravenous drug abusers?

  A: *Staphylococcus aureus*. It is also the most common cause of infective endocarditis in most, including elderly patients.

Q: What valve is usually affected in intravenous drug addicts with endocarditis?

  A: The tricuspid (endocarditis is usually right-sided in intravenous drug addicts)

Q: What organisms most commonly cause subacute bacterial endocarditis?

  A: Viridans group streptococci. This group is probably the second most common cause of bacterial endocarditis. Most of these bacteria are in the oral cavity and are alpha-hemolytic. Examples include *S. mutans*, *S. salivarius*, and *S. mitis*. There is *no* such organism as *Streptococcus viridans*.

Q: Prosthetic valve endocarditis is most often due to what organism?

  A: *Staphylococcus epidermidis* (coagulase-negative staph)

Q: A 28-year-old office worker in a telecommunications company in Omaha, Nebraska had a positive PPD skin test (16-mm). He had a normal chest x-ray and no symptoms. No risk factors were identified. What is the treatment?

  A: This person has latent tuberculosis (TB). If a person has a positive skin test reaction and they are in a low-risk group, preventive therapy with isoniazid or rifampin (different treatment durations) should be initiated. The risk of latent TB progressing to active TB is highest in HIV-positive individuals.

Q: What is the most common cause of neonatal meningitis?

  A: *Escherichia coli* and *Streptococcus agalactiae* (group B strep) are usually listed together as the most common causes. *S. agalactiae* is probably slightly more common than *E. coli*. Infants become infected as they pass through the birth canal.

Q: What are the most common bacteria that you would try to culture from patients presenting with food-associated diarrhea?

  A: *Salmonella*, *Shigella*, and *Campylobacter*. *Campylobacter* is not in the family Enterobacteriaceae as are *Salmonella* and *Shigella*.

Q: Diarrhea accompanied by fever, cramps, and blood and mucus in the stools is associated with infections due to what organisms?

  A: *Shigella* (all species), some strains of *E. coli* (e.g., enterohemorrhagic and enteroinvasive strains), some cases of salmonellosis, *Clostridium difficile*, and the protozoan *Entamoeba histolytica*. Most other agents of gastroenteritis, such as *Clostridium perfringens*, enterotoxigenic *Escherichia coli*, *Vibrio cholerae* and viruses, cause a watery diarrhea.

Q: What part of immunity is not specific and has no memory?

  A: Natural or innate immunity. It consists of macrophages, natural killer cells, mast cells, eosinophils, basophils, the inflammatory response, complement, Toll-like receptors, and so forth. It is rapid response.

Q: With respect to the previous question, lactoferrin and transferrin can be considered part of innate immunity. Why?

  A: Lactoferrin and transferrin bind iron for use by human cells. Bacteria, as well as all other living organisms, also require iron. Lactoferrin and transferrin therefore serve to sequester the body's iron so it is more difficult for invading bacteria to have access to it. To bypass this part of innate immunity, many pathogenic bacteria secrete iron-binding proteins, called siderophores, that allow them to get iron and therefore survive in the host.

Q: Describe the role of Toll-like receptors in innate immunity.

A: Toll-like receptors belong to a class of cell-associated proteins called pattern recognition receptors (PRRs). PRRs are proteins in mammals that function to recognize pathogen-associated molecular patterns (PAMPs). PAMPs are microbe specific molecular structures. An example of a PAMP is the lipopolysaccharide (LPS) of gram-negative bacteria. PRRs such as Toll-like receptors (and others) activate immune cells causing an induction of innate and adaptive immunity. For example, after infection by a microbe, PRRs activate the complement pathway (innate immunity) and can induce the production of a variety of cytokines (e.g., interleukin [IL]-1, IL-6, and tumor necrosis factor [TNF]) and other immune mediators (e.g., chemokines) to induce the inflammatory response, recruit neutrophils to the infection site, and activate macrophages to phagocytize and kill the pathogens.

Q: What virus is responsible for most of the viral diarrheas in infants younger than 2 years of age?

A: Rotavirus accounts for 40–60% of acute gastroenteritis in children under the age of 2. By age 4 over 90% of American children have humoral antibodies suggesting a high rate of inapparent infection in early life. Norovirus commonly causes outbreaks of diarrhea among older children and adults (number one viral cause).

Q: Some bacteria, such as *Neisseria gonorrhoeae*, undergo antigenic variation (also called phase variation) of the pili (fimbriae) or outer membrane proteins. What are the consequences of such antigenic or phase variation?

A: It allows the bacteria to evade immune surveillance and may result in persistent infection or reinfection with the same organism. This or analogous types of antigenic variation occur in other bacterial (relapsing fever) and in some parasitic (African trypanosomiasis, malaria) infections. A similar phenomenon in viruses, such as influenza, is referred to as antigenic drift.

Q: Why is there frequently an association between flooding and cholera outbreaks in developing countries?

A: *Vibrio cholerae* is a strict human pathogen that is spread by the fecal–oral route. Flooding causes contamination of local water supplies.

Q: What is C-reactive protein?

A: C-reactive protein, or CRP, is one of several acute-phase proteins found in plasma that are markers for inflammation. The acute phase response is relatively nonspecific and consists of vasodilation, platelet aggregation, neutrophil chemotaxis, fever, leukocytosis, and so on. It occurs in such widely diverse conditions as infection, allergic reactions (e.g., rejection after renal transplantation), thermal and hypoxic injuries, trauma, surgery, and malignancies (e.g., lymphoma and sarcoma). IL-1, IL-6 and other cytokines induce CRP synthesis by hepatocytes in the liver. A normal CRP is unlikely in bacterial infections. A very high CRP (>100 mg/L) is more likely to occur in bacterial rather than viral infections. Intermediate CRP levels (10–50 mg/L) can be seen in both bacterial and viral infections. Measuring CRP levels can be useful in monitoring treatment.

Trivia: The name for this protein is derived from its ability to react with C-polysaccharide, a component of the cell wall of *Streptococcus pneumoniae*.

Q: What is the major predisposing factor for acquiring legionellosis?

A: Immunosuppression. The typical patient is an older immunosuppressed person, such as an individual who smokes, is a moderate to heavy user of alcoholic beverages, may have chronic pulmonary disease, and may be immunosuppressed (due to other disease or immunosuppressive drugs). It has been found in hospital plumbing systems as well as in long-term care facilities. Infection is probably underreported.

Q: What is the most common cause of septic arthritis in young adults (aged 18–30)?

A: *Neisseria gonorrhoeae*

Q: What stage or stages of syphilis are infectious?

A: The chancre of primary syphilis and the papular rash of secondary syphilis are infectious.

Q: List three ways that acquired immunity can be achieved by immunization.

A: Active immunization, passive immunization, and adoptive transfer

Q: List the major alpha- and beta-hemolytic streptococci.

A: Alpha: *S. pneumoniae* and the viridans group. Beta: *S. pyogenes* (always) and *S. agalactiae* (usually). There are other beta-hemolytic streptococci, but you do not need to learn them. Most other streptococci show variable hemolytic patterns. Make sure you learn the genus and species of the streptococci for standardized exams. They commonly do not use group names, for example, *Streptococcus pyogenes* instead of group A and *Streptococcus agalactiae* instead of group B. You rarely need to know the hemolytic patterns for other bacteria.

Q: What is the major virulence factor of *Streptococcus pneumoniae*?

A: The antiphagocytic polysaccharide capsule

Q: What causes Weil disease?

A: *Leptospira interrogans*. Weil disease is an uncommon severe form of leptospirosis, with icterus (jaundice), marked hepatic and renal involvement, and hemorrhagic diathesis (tendency to bleed after minor trauma) being the main features. The term "Weil (or Weil's) disease" is commonly used incorrectly for leptospirosis in any of its manifestations.

Q: A 7-year-old male patient presented with acute onset of fever and spreading erythema and edema on his right cheek. The involved skin was warm, slightly raised, and the edge was sharply demarcated from uninvolved skin. What is the drug of choice for the organism involved?

A: Penicillin (G or V). This patient has erysipelas (aka Saint Anthony's fire) and the most common cause is *Streptococcus pyogenes*. Erysipelas (the word is singular, not plural) of the face should be empirically treated with a penicillinase-resistant antibiotic such as dicloxacillin or nafcillin to cover for possible *Staphylococcus aureus*. There are other etiologies, including some gram-negative bacilli. Erysipelas can affect all parts of the body but is most common on the extremities.

Q: List some opportunistic infections in AIDS.

A: *Pneumocystis jiroveci* (*carinii*), a fungus that looks like a parasite, causes pneumonia. It is one of the major infections. Others (in no particular order) include *Cryptosporidium* gastroenteritis; *Toxoplasma gondii* encephalitis; *Candida* thrush and vaginitis; *Cryptococcus neoformans* meningitis, encephalitis, and pneumonia; *Coccidioides immitis* meningitis; *Histoplasma capsulatum* pneumonia; *Mycobacterium avium* complex (MAC) pneumonia, disseminated disease, and diarrhea; *Mycobacterium tuberculosis* (Mtb) pneumonia and meningitis; reactivation of varicella-zoster virus (shingles), of herpes simplex virus (oral herpes), and of Epstein-Barr virus (EBV; B-cell lymphomas); cytomegalovirus (CMV; retinitis, pneumonia, encephalitis); and human herpesvirus (HHV)-8 (Kaposi sarcoma).

Q: What group of lymphocytes has the intrinsic ability to recognize and destroy some virus-infected cells and some tumor cells?

A: Natural killer cells. Natural killer cell killing is nonspecific; thus these cells are part of innate immunity. They do not need to recognize the antigen–major histocompatibility complex (MHC) class I or II on the target cell. Natural killer cell killing is mediated by cytotoxins.

Q: A fetus that developed a congenital infection would have an increased level of what immunoglobulin?

A: IgM. IgM is the only antibody synthesized by the fetus, so an increased level is indicative of either a congenital or a perinatal infection.

Q: What is the significance of elevated IgM levels relative to other immunoglobulins in serum of an adult?

A: It is indicative of a recent infection, because IgM is the first antibody produced.

Q: What do the following have in common: *Taenia saginata, T. solium, Diphyllobothrium latum,* and *Hymenolepis nana*?

A: They are all tapeworms.

Q: What do the following cells have in common: Kupffer, microglial, mesangial, synovial A, monocytes, and histiocytes?

A: They are all phagocytic (they are macrophages). Kupffer cells are in the liver, microglia are in the brain, mesangial are in the liver, synovial A cells are in joints (synovial B cells secrete synovial fluid and are not related to the B cells of the immune system), monocytes are in the circulation, and histiocytes are in tissues. Each can function as antigen-presenting cells.

Q: What two special molecules on the surface of antigen-presenting cells function in antigen recognition?

A: MHC class I (found on all nucleated cells including antigen-presenting cells; presentation of endogenous antigens such as from neoplastic or virus-infected cells) and MHC class II molecules (found only on antigen-presenting cells; presentation of processed foreign antigens).

Q: What are the smallest free-living organisms?

A: Mycoplasmas. They have no cell walls.

Q: How can one make the diagnosis of primary syphilis?

A: The gold standard is by dark-field microscopy of suspicious lesions (primary chancre or secondary papules), which shows the spirochete. Rapid plasma reagin (RPR) and VDRL (venereal disease research laboratory) are presumptive tests that must be followed up with specific treponemal tests if positive. These confirmatory tests include the *Treponema pallidum* particle agglutination (TP-PA) and the fluorescent treponemal antibody absorption (FTA-ABS) test, which directly detect antibodies to *T. pallidum* in the patient. A recombinant antigen enzyme immunoassay (EIA) is also available as a screening procedure for detection of syphilis specific antibodies. For a positive EIA screen, the use of the RPR test in combination with RPR quantitation can be used as the confirmatory method. This approach is less expensive because it reduces the use of the more expensive FTA testing in most cases where the EIA and RPR are in agreement. Note: You typically do not need to know what any of the above acronyms stand for.

Q: What genus of tick transmits Rocky Mountain spotted fever?

A: Rocky Mountain spotted fever is primarily spread by *Dermacentor* ticks (hard ticks) such as the wood tick and the dog tick.

Q: What are the clinical symptoms of infection due to *C. difficile*?

A: *C. difficile* causes diarrhea and pseudomembranous colitis, usually 4–10 days after starting antibiotic treatment, but symptoms may appear 1–2 weeks after stopping all antibiotic therapy. Symptoms vary widely from mild diarrhea to severe abdominal pain and are usually accompanied by fever and weakness. Diarrhea is watery and usually nonbloody, but up to 10% may have a bloody diarrhea. Complications of pseudomembranous colitis include perforation and peritonitis and may lead to death.

Q: *Clostridium perfringens* is a common cause of food poisoning. What is the mechanism?

A: The bacteria in food are ingested (no preformed toxin). As they sporulate in the gut lumen, they secrete an enterotoxin that causes hypersecretion of water and electrolytes.

Q: One of the major virulence factors of *C. perfringens* is lecithinase. What role does it play?

A: It is the major virulence factor in the cell destruction seen in gas gangrene (although it is not responsible for the gas production). Because most cell membranes contain lecithin, a potent lecithinase can destroy many tissues quite rapidly. Other bacteria, such as *Pseudomonas aeruginosa*, can secrete a lecithinase.

Q: As an aid in the bacteriological diagnosis of gastroenteritis, stools are cultured on differential and/or selective media to facilitate isolation and separation of intestinal flora into broad groups. What medium component is most widely used to differentiate between the two major groups of enteric organisms?

A: Lactose. MacConkey agar is an example of a selective and differential medium in which lactose provides the means for differentiation. None of the members of the genera *Salmonella* and *Shigella* are able to ferment lactose, whereas *Escherichia coli* and many other coliform bacilli ferment lactose. The ability to ferment this sugar divides all enteric bacteria into two broad groups: lactose fermenters (pink to red colonies on MacConkey agar) and nonlactose fermenters (colorless colonies).

Q: What causes Lyme disease and what are the important characteristics of the organism?

A: *Borrelia burgdorferi* is a motile, spiral-shaped organism (all spiral-shaped bacteria are motile). Although it can be cultivated in the laboratory, diagnosis is usually based on signs, symptoms and serology.

Q: An 18-year-old woman with suspected bacterial meningitis was taken to the emergency department where cerebrospinal fluid (CSF) was collected and sent to the laboratory. What would you suspect the glucose and protein concentrations would be relative to their normal concentrations?

A: Glucose lower, protein higher

Q: From the previous question, a portion of the CSF was centrifuged and the pellet (precipitated material) was Gram stained and viewed microscopically. What is the most likely diagnosis if gram-negative diplococci were seen?

A: Meningitis due to *Neisseria meningitidis*

Q: What is an immunogen?

A: It is a substance that can induce an immune response. All immunogens are antigenic. Not all antigens are immunogenic. For example, the LPS of gram-negative bacteria is antigenic but not immunogenic. We do not develop immunity to the LPS, but we do make antibodies against it; the antibodies are just not protective.

Q: That part of an immunogen to which an antibody is made and to which it will bind is called what?

A: Epitope. Immunogens, especially large ones, frequently have several, if not many epitopes. Bacterial cells, for example, probably have hundreds to thousands of epitopes. A hapten, such as (DNP), usually has a single epitope. T cells also respond to epitopes.

Q: How are *Rickettsia, Ehrlichia, Anaplasma* (formerly *Ehrlichia*), and *Coxiella* grown in the laboratory?

A: All are obligate intracellular parasites and must be grown in tissue culture cells. None will grow on agar plates.

Q: What is babesiosis?

A: Babesiosis is an uncommon tick-borne (the same tick that transmits Lyme disease) parasitic disease caused by *Babesia microti*. The parasite most frequently causes disease in the elderly and immunocompromised individuals. Symptoms include jaundice and anemia.

Q: What does *Legionella* look like under the microscope after Gram staining?

A: It stains poorly with the Gram stain and must be visualized with other techniques such as silver impregnation or fluorescence microscopy.

Q: What other disease should be suspected in a patient with *Streptococcus bovis* bacteremia?

A: Cancer, especially colon cancer

Q: What are the three major genera of acid-fast organisms?

A: *Mycobacterium, Nocardia* (partially acid-fast), and *Cryptosporidium* (parasite)

Q: After staining with the acid-fast stain, what color are the above three organisms?

A: Red

Q: Describe the basic molecular structure of IgG, including the results after papain and pepsin digestion (try to draw IgG).

A: IgG is composed of two heavy (H) chains and two light (L) chains with a total molecular weight of about 150,000. Each L chain is covalently linked by a single disulfide bond to one of the H chains. The two H chains (with attached L) are linked to each other with two disulfide bonds. Drawn in its simplest form, it resembles a "Y" with two antigen-binding sites at the top of the Y. This is also referred to as the variable region and theoretically binds to any of the large number of possible antigens (i.e., epitopes) that can exist. On a single IgG molecule, both L chains are of the same type, either kappa or gamma, depending on the specific amino acid sequence. Papain digestion results in three primary fragments: one Fc fragment (composed of two partial strands of the H chains, including one of the two disulfide bonds holding them together; complement fixation is associated with this portion of IgG; it is analogous to the bottom part of the Y) and two Fab fragments (each with one antigen-binding site; the top of the Y). The three fragments fit together at the center "hinge region," which is flexible. Pepsin digestion yields primarily one large fragment called F(ab')$_2$ and several smaller fragments (from the lower stem of the Y). F(ab')$_2$ consists of the top of the Y (thus it has both antigen-binding sites) + the hinge region + the top part of the stem of the Y including one of the stem disulfide bonds that holds the whole thing together.

Q: Explain the terms "isotype," "allotype," and "idiotype" as they apply to immunology.

A: Isotype refers to one of the five classes or types of antibodies. Each type differs in the specific H chain (Fc portion). Allotype refers to specific differences in the amino acid sequences and three-dimensional structure of the Fc portion of each isotype. In other words, you and I both have IgG (an isotype), but my IgG is a little different from yours in the Fc portion. Similarly, idiotype refers to differences in amino acid sequences and three-dimensional structure, but in the variable (antigen-binding) region of specific antibodies directed against a specific antigen. For example, an IgG might be specific for a particular epitope on the surface of *E. coli*. Another IgG might be made against another epitope on the bacterial surface. Each IgG molecule is of the same isotype (both are IgG) and allotype (both were made in the same person so the Fc regions are identical), but the variable regions are different (i.e., they are different idiotypes).

Q: Name the five types (isotypes) of immunoglobulins and their distinguishing characteristics (structure [with reference to IgG], biological activity, location of highest concentration, ability to cross placenta, etc.).

A: IgG is the major soluble antibody in serum. It functions primarily in response to various infections and is involved in such activities as bacterial lysis, complement fixation, opsonization, and toxin inactivation. It is the only immunoglobulin (Ig) that can cross the placenta (but read on). There are four subclasses (IgG$_1$, IgG$_2$, IgG$_3$, and IgG$_4$), each having increased biological activity in some aspects, such as opsonization (IgG$_1$ and IgG$_3$) or passage through the placenta (all but IgG$_2$). A membrane bound form of IgG is found on the surface of plasma cells and memory cells after clonal expansion of B cells.

IgA is the second most common Ig in serum where it exists as a monomer. Serum IgA has a structure similar to that of IgG, but it has a slightly higher molecular weight (about 160,000). There are two subclasses, IgA$_1$ and IgA$_2$. IgA is most important in its secretory form (sIgA) where it exists as a dimer. A peptide called the joining or J chain holds the dimer together. The dimer picks up another peptide, the secretory component, during its secretion. sIgA is found in colostrum, mucus, saliva, tears, etc. At mucosal surfaces, sIgA can interfere with microbial attachment.

IgM is the largest Ig in its typical pentameric form that is found in serum (molecular weight of about 900,000), consisting of five Ig monomers held together with disulfide bonds and a J chain. It is the first antibody produced in response to a foreign antigen. It is also the only antibody synthesized by the fetus. The monomeric form of IgM is found on the surface of B cells along with IgD (they are the B cell receptors). It is an excellent activator of complement.

IgD is found primarily on the surface of B cells, where along with IgM, it functions as a B-cell receptor. It is slightly larger in molecular weight than IgG but otherwise is similar in structure. IgE is lower in serum concentration than any other Ig. It is found primarily on the surfaces of mast cells where it functions in hypersensitivity (e.g., anaphylaxis) and in response to parasitic infections.

Q: What is class (isotype) switching?

A: IgG, IgM, and other Igs are different isotypes because their constant regions are different. For example, a B cell might be making a specific IgM that binds to a particular antigen. Under the influence of various cytokines, this B cell can start making a different class of antibody such as IgG. The specificity of the IgG is the same as was the IgM because there was no change in the variable region that binds to the antigen.

Q: What are teichoic acids?

A: They are cell wall structural components found only in gram-positive bacteria. There are two kinds: lipoteichoic acids (covalently linked to phospholipids in the cytoplasmic membrane) and cell wall teichoic acids (covalently linked to peptidoglycan). Teichoic acids contribute to pharyngeal epithelial cell attachment by *Streptococcus pyogenes* and commonly contribute to virulence (usually in colonization) in other gram-positive bacteria.

Q: What are the most obvious differences between the cell walls of gram-positive and gram-negative bacteria?

A: Gram-positives have a thick cell wall composed primarily of peptidoglycan; gram-negatives have a more complex cell wall consisting of an outer membrane containing the LPS. They have a thin layer of peptidoglycan that is found in the periplasmic space.

Q: What are the common names for the following: *Enterobius vermicularis* and *Trichuris trichiura*?

A: Pinworm and whipworm, respectively

Q: Which one of the previous parasites is most common in the United States?

A: Pinworm, with approximately 40 million infected. Whipworm is not as common in the United States (about 2–3 million infected, mostly southeastern states) but is estimated to infect 25% of the world's population.

Q: What organisms are most likely to cause food poisoning associated with the ingestion of poultry?

A: Most likely *Campylobacter* or *Salmonella*, but it could be other organisms such as *Clostridium perfringens*. Some cases of salmonellosis have been associated with the ingestion of raw eggs. Of course other non-poultry foods can be sources of food poisoning with any of these if they are contaminated.

Q: What is the composition of the DTaP vaccine and the vaccines for *Streptococcus pneumoniae* and *Haemophilus influenzae* type b?

A: **DTaP:** Diphtheria toxoid, tetanus toxoid, and acellular pertussis vaccine (pertussis toxoid plus various cellular components of *Bordetella pertussis*)

Streptococcus pneumoniae: The standard pneumococcal polysaccharide vaccine contains 23 of the most common capsular serotypes and is not effective in children under 2 years of age. For children under 2, the heptavalent pneumococcal conjugate vaccine is used (seven capsular serotypes).

Haemophilus influenzae type b: Capsular polysaccharide (serotype b). The Hib vaccine is a conjugate vaccine that can have various formulations such as the capsule polysaccharide conjugated to N. meningitidis outer membrane proteins, or tetanus or diphtheria toxoids.

Q: What types of vaccines are measles, mumps, rubella, polio (Salk and Sabin), hepatitis A and B, and rabies?

A: **Measles, mumps, and rubella:** attenuated (live) viruses
Polio: Salk, inactivated (dead) virus; Sabin, attenuated virus
Hepatitis: A, inactivated virus; B, subunit
Rabies: inactivated virus

Q: In bacteria, what are the H, O, and K antigens?

A: Flagella, repeat carbohydrate structure (also called somatic antigen) of the gram-negative LPS, and the capsule, respectively. They are useful for serological identification of bacteria.

Q: To what enzymes do the beta-lactam antibiotics bind?

A: Transpeptidases, also called penicillin binding proteins

Q: What is the bacterial target for lysozyme?

A: Peptidoglycan, specifically the beta 1–4 bond between N-acetylmuramic acid and N-acetylglucosamine. It is important to remember that lysozyme degrades peptidoglycan, and cell wall active antibiotics inhibit its synthesis.

Q: What are antitoxins?

A: Antibodies produced in response to a toxin. These antibodies are protective.

Q: What is a toxoid?

A: It is a toxin that has been altered (chemically, heat, etc.) that inactivates the toxin (it is no longer toxic) but retains its immunogenicity. That is, antibodies produced against the toxoid (e.g., the tetanus toxoid) are protective against the active toxin.

Q: What are porins?

A: Proteins in the outer membrane of gram-negative bacteria (primarily) that are composed of three subunits (trimer) that have a water-filled channel (pore) in the center. They allow only low molecular weight compounds, such as essential nutrients and some antimicrobials, to pass through. For the most part, they form a protective barrier for the bacteria. *Mycobacterium tuberculosis* (acid-fast) and related bacteria also have porins in their cell walls.

Q: What are metachromatic granules?

A: Stored inorganic phosphate in some bacteria that appear red microscopically after staining the bacteria with methylene blue. They can be useful for identifying *Corynebacterium diphtheriae*.

Q: An outbreak of food poisoning occurred at a church picnic held at a city park in July. Most of those affected had symptoms of vomiting and abdominal cramps. A few also experienced diarrhea, but none had fever. All individuals who became ill had eaten inadequately refrigerated potato salad, and they became ill about 1–6 hours after eating the food. What is the most likely cause of this outbreak?

A: *Staphylococcus aureus*

Q: What is unique about *Bacillus* and *Clostridium*?

A: They both produce spores. Both are also gram-positive (not unique). *Bacillus* is an aerobe, and *Clostridium* is an anaerobe.

Q: Name some other genera of pathogenic aerobes besides *Bacillus*.

A: Partial list (you should know these): *Corynebacterium, Mycobacterium, Nocardia, Legionella,* and *Pseudomonas*

Q: Name some other genera of pathogenic anaerobes besides *Clostridium*.

A: Partial list (two you need to know): *Bacteroides* and *Actinomyces*

Q: What is the most common cause of acute infective endocarditis in a previously healthy child without congenital heart disease?

A: *Staphylococcus aureus*

Q: Name two bacteria that can grow at both high and low salt concentrations.

A: *Staphylococcus aureus* (7.5% NaCl) and the enterococci (6.5% NaCl). These are facultative halophiles because they can grow at high or low salt concentrations. Two human pathogens that are true halophiles are *Vibrio vulnificus* (wound infections and gastroenteritis) and *V. parahaemolyticus* (gastroenteritis primarily and wound infections).

Q: What effect do bacteriostatic and bactericidal antimicrobials have on the growth curve of bacteria?

A: Bacteriostatic drugs do not kill; therefore the curve is horizontal after adding the drug. Removal of the drug results in continuation of growth and the curve goes up. Bactericidal antimicrobials kill the cells and the curve goes down after adding the drug. Removal of the antimicrobial has no further effect on the growth curve because the cells are dead.

Q: What are the definitions of MIC and MBC?

A: The MIC, or minimum inhibitory concentration, is the minimum concentration of an antimicrobial agent that inhibits growth of the bacteria. The MBC, or minimum bactericidal concentration, is the minimum concentration of an antimicrobial that is required to kill the bacteria. Because a few bacteria can survive at higher concentrations of the antimicrobial, a reduction of 99.9% of the viable bacteria is used as the endpoint for the MBC.

Q: How do you interpret the results in the agar disk diffusion (Kirby-Bauer) method of susceptibility testing?
A: Three parameters are needed: The identification of the organism, the antibiotic being tested, and the diameter of the zone of inhibition around the antibiotic disk. The results are read by comparing the diameter of the zone to zone diameters that are provided in standardized tables. Results are qualitative and reported as susceptible, resistant, or, in some instances, intermediate.

Q: Compare and contrast exotoxins and endotoxins (make a table).
A:

Table 15-1

	Exotoxin	Endotoxin
Cell type	Gram positive and negative	Gram negative
Production	Secretion	Cell lysis
Pyrogenic	No*	Yes
Composition	Protein	LPS
Heat labile	Yes†	No
Induces antitoxin	Yes	No
Converted to toxoid	Yes	No
Toxicity‡	High	Low
Specificity§	High*	Low
Shock	No*	Yes

* Superantigens are exceptions.

† The heat stable toxin of *E. coli* and the staphylococcal enterotoxin are heat stable.

‡ Classic exotoxins, such as the botulism or tetanus toxins, are highly toxic, which means only a tiny amount is required to have a significant biological effect on the host.

§ Classic exotoxins are highly specific, which means they have a narrowly defined specific target within a host cell, for example, the diphtheria toxin causes inactivation of elongation factor 2 in protein synthesis. Endotoxins and superantigens are nonspecific, affecting the host by causing cytokine release (e.g., IL-1 and TNF).

Q: What are the standard sterilizing conditions of an autoclave?
A: Fifteen-minute minimum time exposure at 121°C and 15 pounds per square inch pressure. The function of the pressure is to ensure the temperature of the steam reaches 121°C, because nonpressurized steam at sea level is only 100°C.

Q: In a laboratory experiment students were asked to spread 0.1 mL of pasteurized milk across the surface of a blood agar plate. After 48 hours at 37°C, several different kinds of bacterial colonies were observed on the plates. Why were bacteria present, since the milk was pasteurized?
A: Pasteurization does not sterilize. The time and temperature has been set only to kill possible human pathogens found in dairy or other products.

Q: Describe the radiation treatments used to sterilize food that has been irradiated. What is the mechanism of microbial death?
A: Ionizing radiation (gamma or x-rays and accelerated electron beam) forms free radicals from water in any organisms (bacteria, viruses, parasites, insects, etc.) that might be in food. Death is due to damage to nucleic acids and other cellular components by the free radicals.

Q: Ultraviolet (UV) light cannot be used for sterilization. Why not? What is the mechanism of UV light killing?
A: UV light is low energy and cannot penetrate well. It therefore can only be used for surface or air-borne droplet disinfection (e.g., not the interior of a catheter). It kills by causing irreversible DNA damage due to the formation of thymine dimers in the DNA.

Q: What is the mechanism of action of soap against bacteria?
A: Soap is an anionic surfactant and can cause membrane damage. However, it is not too effective as a disinfectant because it does not reach the cell surface well (it has a negative charge as does the surface of bacteria). Its primary use is as a cleansing agent and reduces the population of microbes on surfaces by washing them away.

Q: What is the mechanism of microbial killing by iodine?

A: Iodine kills by two mechanisms: It irreversibly binds to tyrosine, thus inactivating essential enzymes or other proteins, and it is an oxidizing agent. As an oxidizing agent, it causes the formation of disulfide bonds between two sulfhydryl groups in proteins.

Q: In what form is ethylene oxide used and what is the mechanism of its antimicrobial action? How effective is it (i.e., -static or -cidal)?

A: Ethylene oxide is an alkylating agent used as a gas for sterilization (-cidal). Alkylation affects amino and hydroxyl groups on proteins and nucleic acids. Other alkylating agents are glutaraldehyde and formaldehyde that are used in aqueous solution and beta-propiolactone, which is also a gas.

Q: List some diseases that are superantigen mediated and name the superantigen.

A: Staphylococcal toxic shock syndrome: toxic shock syndrome toxin-1
Staphylococcal food poisoning: enterotoxin
Scarlet fever and streptococcal toxic shock-like syndrome: pyrogenic (erythrogenic) toxins
There are other examples, including many virus proteins and some mycoplasmas.

Q: What is the receptor for superantigens and what happens after binding?

A: Superantigens bind to the outside of MHC class II proteins on antigen-presenting cells and to the variable portion of the beta chain on the T-cell receptor of T cells. As a result, T cells release large amounts of cytokines, especially IL-1 and TNF, often resulting in significant biological consequences as in the diseases from the previous question. Even small amounts of superantigen can lead to fever, shock, and death.

Q: What are the biological effects that LPS (endotoxin) has in an infected person and how are these effects mediated?

A: Most of the biological effects of LPS are mediated by cytokines, especially IL-1 and TNF, which are released by macrophages and several other cell types. Effects include, but are not limited to, complement activation, the stimulation of B cells, fever, shock, disseminated intravascular coagulation, and death. The biological effects of LPS and superantigens are analogous.

Q: What causes Pontiac fever?

A: *Legionella pneumophila*. It is a mild "walking" pneumonia.

Q: What is triclosan?

A: It is a chemical biocide used in many antibacterial soaps, shampoos, and mouthwashes. It can also be infused into household products (cutting boards, sinks, toys, etc.) to make them antibacterial. There is some evidence that some bacteria are becoming resistant.

Q: What are some of the important normal flora organisms at each of the following sites and why are they important: blood, central nervous system (CNS), skin, eye, mouth, nose and nasopharynx, stomach, small intestine, colon, vagina, urethra, and urinary bladder?

A:

- **Blood, CNS, and bladder:** These and other tissues and organs are normally sterile. There can be a transient bacteremia occasionally, such as after brushing the teeth. If the urine in the bladder has bacteria, the person has a urinary tract infection (UTI).
- **Skin:** *Staphylococcus aureus* and *S. epidermidis* (involved in various skin, opportunistic, and nosocomial infections), *Propionibacterium acnes* (anaerobe living in sebaceous glands and hair follicles; contributes to the inflammation of acne)
- **Eye:** *Haemophilus, Moraxella, Streptococcus* (can all be involved in conjunctivitis)
- **Mouth:** Mostly anaerobes living in the gingival crevice. Can be involved in aspiration pneumonia (as can the contents of the stomach), soft tissue infections (gingivitis, periodontal disease, actinomycoses, etc.), infective endocarditis (viridans strep). The yeast *Candida albicans* (oral thrush) is also found here and elsewhere.
- **Nose and nasopharynx:** *Streptococcus pneumoniae, Haemophilus influenzae, Neisseria meningitidis* (all can be causes of meningitis and other infections), *Staphylococcus aureus* (you should know disease associations), *Mycoplasma pneumoniae* (atypical pneumonia and other respiratory infections), various anaerobes (soft tissue and other infections).
- **Stomach:** *Helicobacter pylori* (ulcers, stomach cancer)
- **Small intestine:** mixed anaerobes

- **Large intestine:** Primarily anaerobes, including *Bacteroides fragilis* (and related *Bacteroides*) and *Clostridium*. Other flora consists of *Enterococcus*, members of the family Enterobacteriaceae (e.g., *E. coli*, *Proteus*, *Enterobacter*, etc.), and *Candida* (and other yeasts).
- **Vagina:** *Lactobacillus* (plays role of maintaining normal low vaginal pH that is too acidic [pH < 4.5] to allow for colonization by many disease-causing microbes; hydrogen peroxide production by the normal lactobacilli is also protective), *E. coli* and *Streptococcus agalactiae* (both are important causes of neonatal sepsis and meningitis), *Gardnerella vaginalis* (associated with bacterial vaginosis, as are other bacteria such a *Mobiluncus*), and *Candida* (vulvovaginitis)
- **Urethra:** *E. coli* (number one cause of UTIs)

Note that the above list is partial. There are many others.

Q: An increase in neutrophils indicates infection with what broad category of organisms (bacteria, viruses, or parasites)? Increase in lymphocytes? Increase in eosinophils?

A: Neutrophils: bacteria; lymphocytes: viruses; eosinophils: parasites. There are, of course, exceptions.

Q: In a patient with an asymptomatic UTI, how many bacteria (measured in the standard colony forming units, or cfu) must be present in 1 mL of urine for a positive diagnosis of UTI?

A: 100,000 or 10^5 cfu/mL

Q: A 27-year-old woman presented with dysuria and pyuria. Routine bacteriological culture did not show growth of any organisms after 48 hours. Assuming this is a bacterial infection, what are some possible causes?

A: Consideration should be given to organisms that do not grow on standard laboratory media, such as blood agar plates or MacConkey agar, or organisms that have a long generation time. *Chlamydia* grows within cells such as in tissue culture, but not on agar plates. *Mycobacterium* not only has a long generation time (meaning it takes prolonged incubation before colonies are observed) but also does not grow well on typical laboratory media that is normally used for routine culture.

Q: What are extended-spectrum beta-lactamases?

A: Plasmid-borne enzymes produced by gram-negative bacteria that mediate resistance to extended-spectrum (third-generation) cephalosporins (e.g., ceftazidime, cefotaxime, and ceftriaxone) and monobactams (e.g., aztreonam). They do not affect cephamycins (e.g., cefoxitin and cefotetan) or carbapenems (e.g., meropenem or imipenem).

Q: What are the clinical manifestations of disseminated gonococcal infection?

A: Hematogenous dissemination of *Neisseria gonorrhoeae* occurs in up to 3% of asymptomatic primary infections in women. Symptoms start 7–28 days after infection and within seven days of the last menstrual cycle. The most common signs are arthritis, polyarthralgia, tenosynovitis, and a papular rash. Most patients do not have genitourinary symptoms; however, up to 80% of cervical cultures will be positive for *N. gonorrhoeae*.

Q: What causes syphilis and what are the important characteristics of the organism?

A: *Treponema pallidum*. It is spiral-shaped and motile (all spirochetes are motile). It cannot be grown on laboratory media but will grow in some animals such as rabbits.

Q: What are the growth requirements for *Haemophilus influenzae*?

A: *H. influenzae* requires the addition of hematin (X factor) and NAD (V) factor to laboratory growth media.

Q: What effect does bacterial meningitis usually have on the glucose level of CSF?

A: Decrease. This is also commonly true for fungi.

Q: What age group is at the highest risk for *Haemophilus influenzae* infection?

A: Young children from about 3 months to 3 years. However, in the elderly, as their immune system becomes less effective, the infection rate increases again. The use of the Hib vaccine has decreased the infection rate in young children.

Q: What is the number one cause of bacterial meningitis in adults?

A: *Streptococcus pneumoniae* is the number 1 cause of bacterial meningitis in all age groups, including 3 months to 3 years.

Q: A specimen container from a patient was labeled as "sputum." When the laboratory technician observed the specimen microscopically, he counted an average of 18 epithelial cells per low power field (100×, total magnification). What should be done next?

A: Request a new specimen. Sputum specimens should not have more than 10 squamous epithelial cells per low power field. More than 10 indicates saliva (the epithelial cells are probably cheek cells).

Q: What is the number one cause of bacterial pneumonia in adults?

A: *Streptococcus pneumoniae*

Q: Two second-year medical students were preparing for the next microbiology exam that included a section on diarrheal diseases. A case they were reading discussed a patient with an infectious bloody diarrhea. One of the students thought this was a case of bacterial diarrhea and the other student thought it was viral diarrhea. Who was right?

A: This is most likely a case of bacterial diarrhea. Viral diarrheas are usually watery with no blood. Bloody diarrhea can also be caused by some parasites and can have noninfectious causes.

Q: Which important pathogenic bacteria are isolated on the following media: Thayer-Martin, Tellurite, buffered charcoal yeast extract, Bordet-Gengou, chocolate, MacConkey, mannitol salt, Lowenstein-Jensen, Middlebrook, Sabouraud dextrose?

A: Most of these media are selective for certain species or groups of bacteria; however, as is true for most selective media, other bacteria may grow, usually just not as well as the bacteria that are being selected. Most selective media are also differential.

- **Thayer-Martin agar:** *Neisseria.* You will see this medium referred to as modified Thayer-Martin or Thayer-Martin modified. Do not worry about the modifications! There are too many to keep track of. This selective medium allows most strains of *Neisseria* to grow (a few may be inhibited) while inhibiting most other gram negatives and gram positives. You do not need to memorize the components. Just know that there are antibiotics that inhibit most other bacteria.
- **Tellurite agar:** *Corynebacterium diphtheriae.* Tinsdale agar is another medium that can be used for the selection and differentiation of *C. diphtheriae.*
- **Buffered charcoal yeast extract agar:** *Legionella.* This medium is supplemented with cysteine and iron (ferric pyrophosphate).
- **Bordet-Gengou agar:** *Bordetella.* This medium is supplemented with sheep blood.
- **Chocolate agar:** *Neisseria* and (if supplemented with the X and V factors) *Haemophilus.* Chocolate agar contains no chocolate. It turns brown and looks like chocolate after blood is added to the agar base while it is still hot (about 80°C), resulting in lysis of the erythrocytes. It is a nonselective medium that supports the growth of many bacteria. Blood agar (which is also nonselective) is made by adding blood (usually from sheep) to melted agar base at a cooler temperature (about 45–50°C).
- **MacConkey agar:** Gram-negative bacteria, particularly those in the family Enterobacteriaceae. It is also differential in that it can differentiate between lactose fermenters (e.g., *E. coli* and *Klebsiella*; red/pink colonies) and lactose nonfermenters (e.g., *Salmonella* and *Shigella*; colorless colonies).
- **Mannitol salt agar:** *Staphylococcus.* Bacteria that cannot grow at high salt concentrations (7.5%) will not grow. The medium also differentiates between the staphylococci that ferment mannitol (*S. aureus*) and those that do not (most coagulase-negative staphylococci).
- **Lowenstein-Jensen medium:** *Mycobacterium* (most species). This medium is green because of the dye (malachite green) that inhibits many gram positives.
- **Middlebrook agar:** *Mycobacterium.* There are also many modifications of this medium, some selective, some not.
- **Sabouraud dextrose agar:** selective for yeasts and fungi

Q: In a microbiology laboratory for medical students, one of the students was observing a Gram-stained slide of his unknown bacteria. He could not decide if the bacteria were gram-positive clusters or chains. To help him determine if his unknown was *Staphylococcus* or *Streptococcus*, the lab instructor gave him a small reagent bottle. What was in the bottle?

A: Hydrogen peroxide. *Staphylococcus* is catalase positive (bubbles when mixed with hydrogen peroxide) and *Streptococcus* is catalase negative.

Q: In the previous question, the student added a drop of hydrogen peroxide to some cells of his unknown bacteria that he had placed on a glass microscope slide. The bubbles (oxygen) indicated that his unknown was a *Staphylococcus*, but which one? To help answer this question, the lab assistant gave him a small tube containing a clear fluid. What was the fluid?

A: Rabbit plasma. *Staphylococcus aureus* incubated in rabbit plasma causes coagulation (i.e., coagulase positive). If the plasma stays liquid, the organism is coagulase negative (other staphylococci).

Q: What is the drug of choice for methicillin-resistant *Staphylococcus aureus*?

A: Vancomycin (with or without gentamicin and/or rifampin). This is also the drug of choice for methicillin-resistant *Staphylococcus epidermidis*.

Q: What is the time of onset of the following poststreptococcal (*S. pyogenes*; group A strep) complications and what is the usual precipitating streptococcal infection: scarlet fever, rheumatic fever, and glomerulonephritis?

A: **Scarlet fever:** 1–2 days after the first symptoms of pharyngitis
Rheumatic fever: 1–6 weeks after pharyngitis
Glomerulonephritis: 1–4 weeks after pharyngitis or skin infection

Q: What is M protein?

A: It is a major component (along with peptidoglycan and teichoic acid) of the cell wall of *Streptococcus pyogenes*. It is antiphagocytic and can function in adhesion to host cells and as a superantigen. Another name for the M protein is fimbriae (in streptococcus). The M protein can be visualized in the electron microscope as hair-like fibers on the outer surface of the cells. They are composed of two alpha-helical chains. There are over 80 serotypes of *S. pyogenes* based on different M proteins. Antibody to the M protein can cross-react with heart tissue and may result in rheumatic fever.

Q: What is the "spreading factor" of *Streptococcus pyogenes*?

A: Hyaluronidase

Q: What is the mode of action of the diphtheria toxin?

A: It inhibits protein synthesis by causing an ADP ribosylation of elongation factor 2.

Q: What is the mode of action of exotoxin A of *Pseudomonas aeruginosa*?

A: Although exotoxin A has no relationship to the diphtheria toxin, their modes of action are identical (see previous question).

Q: A 19-year-old man presented to his family physician with a furuncle on the back of his neck. What would a Gram stain of an aspirate show?

A: Gram-positive cocci in clusters (*Staphylococcus aureus* is the usual cause)

Q: ASO titers are useful for diagnosis of what diseases?

A: Acute rheumatic fever and poststreptococcal glomerulonephritis
Note: You typically are not responsible for how the ASO titers are determined, but a little background might make it easier to learn: The patient's serum is serially diluted and the titer of ASO antibodies is determined. A titer of 200 Todd units (you do not need to know what a Todd unit is) or greater is useful in diagnosis. Thus knowledge of the previous titer before the patient became symptomatic is not required.

Q: Briefly describe the mechanisms by which *Streptococcus pyogenes* causes acute rheumatic fever and poststreptococcal glomerulonephritis.

A: Rheumatic fever is due to cross-reacting antibodies to the M protein of *S. pyogenes* and cardiac myosin. Glomerulonephritis is due to the deposition of immune complexes on the kidney basement membrane.

Q: What three bacterial food poisonings are rapid onset and short duration? How do each present?

A: *Staphylococcus aureus*, presents as vomiting; *Clostridium perfringens*, diarrhea; and *Bacillus cereus*, vomiting or diarrhea, depending on the strain.

Q: What is the most common cause of osteomyelitis?

A: *Staphylococcus aureus*

Q: What is the most common cause of osteomyelitis in sickle cell disease?

A: The answer to this question is controversial and depends on which reference you believe. Although patients with sickle cell disease are more prone to osteomyelitis due to *Salmonella*, the most common cause in the United States is still *Staphylococcus aureus*. In other parts of the world, such as Africa and India, *Salmonella* may be the most common cause. Just to be controversial, there are also references supporting *Salmonella* as the most common cause in the United States.

Q: A patient presenting with urethritis, conjunctivitis, uveitis, and arthritis may have a current or prior infection with what organism?

A: The patient is most likely a male with a genital infection due to *Chlamydia trachomatis*.

Q: A 9-year-old girl presented to her pediatrician in Bellevue, Nebraska with a mildly pruritic erythematous rash on her right leg. It soon developed a vesiculopustular appearance with an erythematous base. Several days later she noted a ring of erythema developing around the site. The ring continued to enlarge for about one week. The girl had been camping in Minnesota with her family about two weeks before her first visit. There were numerous insects, as well as ticks, seen during the camping trip. The patient recalled some itching in the same area of her right leg, which she scratched through her jeans. There was no tick seen at the site of itching, but she removed one from her hair. She had numerous mosquito bites as well. She was most likely infected with what organism?

A: *Borrelia burgdorferi*, the cause of Lyme disease. Lyme disease is not endemic in Nebraska. It is not unusual for vacationers to be diagnosed with Lyme disease sometime after visiting an endemic area. The rash is commonly described as having a characteristic "bull's-eye" appearance with central clearing, but this is not as common as the rash being homogeneously red.

Q: The time of onset of symptoms of food poisoning varies depending on the etiological agent and can be useful in diagnosis. List three bacteria that typically result in symptoms in less than one day.

A: *Staphylococcus aureus* (produces an enterotoxin in contaminated food), *Clostridium perfringens* (due to ingestion of the organism which produces an enterotoxin during sporulation in the intestinal tract), and *Bacillus cereus* (produces an enterotoxin in contaminated food, such as reheated rice and other foods)

Q: What bacterium produces a superantigen that is responsible for the resulting food poisoning?

A: *Staphylococcus aureus*. The enterotoxin produced in food is a superantigen and is identical to the toxic shock syndrome toxin-1 produced by this organism. Although this enterotoxin is also a superantigen, it acts as a neurotoxin in the gastrointestinal (GI) tract, not as a superantigen. As a neurotoxin, it induces vomiting.

Q: Eosinophilia is suggestive of infection with what group of organisms?

A: Parasites, particularly the helminths (worms)

Q: What are some infectious diseases carried by ticks in the United States?

A: Rocky Mountain spotted fever, Lyme disease, human monocytic ehrlichiosis (*Ehrlichia chaffeensis*), human granulocytic anaplasmosis (previously human granulocytic ehrlichiosis; *Anaplasma phagocytophilum*), tularemia (can also be spread other ways), and babesiosis (*Babesia microti/divergens*)

Q: What genus of tick is the carrier of Lyme disease in the United States?

A: *Ixodes*

Q: What is the mode of action of the sulfonamides?

A: Sulfonamides are metabolic analogs of para-aminobenzoic acid. They competitively inhibit the binding of para-aminobenzoic acid to the enzyme dihydropteroate synthetase, which is required for the bacterial synthesis of tetrahydrofolic acid.

Q: A gram-positive, catalase-negative coccus was isolated from a patient with subacute bacterial endocarditis. On a blood agar plate, the organism was resistant to optochin. What hemolytic reaction would be expected?

A: Alpha-hemolysis because the organism is most likely a viridans streptococcus

Q: *E. coli* O157:H7 is a strain that can cause food poisoning with sometimes serious complications, such as hemolytic uremic syndrome. In the serological designation of this strain, the "H" refers to what?

A: Flagella

Q: Which food-borne pathogen is an important cause of bacterial meningitis in both neonates and immunocompromised adults?

A: *Listeria monocytogenes*

Q: In an enteric pathogen screen, the clinical lab usually is looking for three major organisms. What are they?

A: *Salmonella*, *Shigella*, and *Campylobacter*

Q: For which of these three (previous question) is lactose negative (non-fermenter) an important characteristic?

A: *Salmonella* and *Shigella*. It helps to differentiate these two from *E. coli*, which is normal flora.

Q: Of the two that are lactose negative, which one is nonmotile?

A: *Shigella*. It is also hydrogen sulfide negative. *Salmonella* is H_2S positive.

Q: Salmonellosis is normally not treated with antibiotics. Why?

A: The disease normally resolves without antimicrobial intervention. The use of antibiotics can also select for antibiotic resistant strains of *Salmonella* and normal flora bacteria.

Q: What is the endemic area in the United States for the plague?

A: Four corners (New Mexico, Arizona, Colorado, Utah) area especially, but southwest United States in general.

Q: A 57-year-old male bus driver who smoked heavily developed a pneumonia in which he produced a thick sputum that became bloody later in the infection. A gram-negative bacillus, which produced extremely mucoid colonies on blood agar plates, was isolated. The organism was nonmotile and urease positive. What is the major virulence factor of the most likely pathogen?

A: The most likely organism is *Klebsiella pneumoniae*. The antiphagocytic capsule is the major virulence factor. The capsule is responsible for the mucoid colonies and the thick sputum.

Q: What is the shape of the bacteria that cause Weil disease?

A: They are spiral shaped. Weil disease is a rare but serious manifestation of leptospirosis.

Q: What is the name of the organism that causes trichomoniasis?

A: *Trichomonas vaginalis*

Q: From the previous question, how is this organism classified?

A: It is a parasite that is a protozoan. Within the protozoans, it is a flagellate (it has flagella).

Q: A 2-week-old low-birth-weight male infant presents with symptoms consistent with bacterial meningitis. What are the most likely infecting organisms?

A: *Escherichia coli* and *Streptococcus agalactiae* are the two most common causes of neonatal bacterial meningitis.

Q: A 23-year-old woman presented to her physician with symptoms consistent with a UTI. Urine culture yielded a gram-negative bacillus that swarmed across the surface of the blood agar plate. What is the most likely organism?

A: *Proteus*

Q: Food poisoning due to what organism is the result of enterotoxin secretion in the colon during sporulation?

A: *Clostridium perfringens*

Q: What is the drug of choice for trichomoniasis?

A: Metronidazole, which should be given to both partners, because this is a sexually transmitted disease (STD).

Q: What is responsible for a UTI in which the isolated organism is a gram-negative bacillus that is oxidase positive and that produces a bluish-green pigment?

A: *Pseudomonas aeruginosa*. UTIs due to *P. aeruginosa* are usually catheter associated. The water-soluble pigments are pyocyanin (blue) and fluorescein (yellow).

Q: What is the most common cause of pneumonia in a patient with cystic fibrosis?

A: *Pseudomonas aeruginosa* is the most common, but initial infection is often due to *Staphylococcus aureus*, especially in younger patients. *P. aeruginosa* strains in cystic fibrosis have a mucoid antiphagocytic polysaccharide capsule, making treatment difficult because antibiotics cannot penetrate into the cells well. *Haemophilus influenzae* is also commonly isolated. Other organisms that can be involved include *Burkholderia cepacia*, *Stenotrophomonas maltophilia*, *Alcaligenes* spp., *Aspergillus* spp., non-TB mycobacteria, and respiratory viruses.

Q: What is the most common cause of diarrhea after long-term antibiotic administration?

A: *Clostridium difficile* diarrhea often begins 3–4 days (as few as two days) after antibiotic administration. Pseudomembranous colitis may result. A hypervirulent strain of *C. difficile* appears to be emerging as are community-associated disease and disease associated with the use of fluoroquinolones.

Q: Why does *Clostridium perfringens* generally affect only traumatized tissue?

A: *C. perfringens*, as a typical anaerobe, requires a low oxidation-reduction (redox) potential to grow. Nontraumatized tissue generally has a high redox potential, making it difficult for *C. perfringens* to grow.

Q: What are the two most important clostridial diseases that can be treated with antitoxin?

A: Antitoxin is particularly important for treatment of disease due to *C. tetani* and *C. botulinum*. The antitoxin is only effective early in the course of the disease because once the toxin is bound to receptor cells, the antitoxin is ineffective.

Q: How should specimens from suspected anaerobic infections be transported to the laboratory?

A: Because oxygen can be lethal to anaerobic organisms, specimens must be transported in an anaerobic transport tube.

Q: What important infectious disease is characterized by rice-water stools?

A: Cholera

Q: When should serologic tests be performed for syphilis in a pregnant patient?

A: During the first and third trimesters of pregnancy. Because the fetus is susceptible to infection after the fourth month, it is important to diagnose and treat early in the pregnancy. Testing in the last trimester is particularly important in those patients at high risk of acquiring syphilis. Congenital manifestations of syphilis include stillbirth, abortion, and birth defects. *Treponema pallidum* penetrates the maternal fetal barrier infecting the fetus in utero.

Q: What is the major virulence factor of *Helicobacter pylori*?

A: The most important for you to know about is urease, which allows the organism to split urea into ammonia and carbon dioxide. The ammonia (alkaline pH in solution) neutralizes the acidity of the stomach, allowing the organism to survive. *H. pylori* has other important virulence factors, such as outer inflammatory protein A, vacuolating cytotoxin, cytotoxin-associated antigen, and colonization factors.

Q: What infection is due to *Mycobacterium marinum*?

A: Skin infection, usually a self-limiting granuloma, associated with water (e.g., fishermen, swimming pools, aquarium workers/hobbyists).

Q: How is *Mycobacterium tuberculosis* spread in the human body to extrapleural sites?

A: Hematogenously, via draining lymph vessels and nodes, or by direct erosion to pulmonary veins

Q: The Ghon complex is associated with what stage of tuberculosis?

A: The Ghon (or primary) complex (visible by x-rays) develops as the initial tubercles (not visible with x-rays) calcify in primary tuberculosis. The calcified lesion plus the associated enlarged lymph node is the Ghon complex.

Q: How is Hansen disease diagnosed?

A: Depending on the type (tuberculoid, lepromatous, or borderline), Hansen disease (leprosy) is diagnosed by a combination of clinical signs and symptoms, microscopic visualization of the organisms in tissue specimens or scrapings, and demonstration of granulomas. The organism cannot be cultured in vitro.

Q: In Hansen disease, what is the major immune response involved in the progression of the disease?

A: Cell-mediated immunity. A weak cell-mediated immunity correlates with the most severe form of the disease.

Q: How does *Francisella tularensis* infection occur?

A: Tularemia is a zoonotic disease spread to hunters, veterinarians, or others who are exposed to small wild mammals such as rabbits (rabbit fever). Infection can occur by direct contact (via skin abrasions or contamination of the conjunctiva), by eating improperly cooked food, by inhalation of contaminated droplets, or by bites from ticks, deerflies, or mosquitoes. *Francisella tularensis* is a Centers for Disease Control and Prevention Category A agent of bioterrorism.

Q: How is yersiniosis spread?

A: By contaminated water and food. The etiological agents are *Yersinia enterocolitica* (children younger than 1 year of age) and *Yersinia pseudotuberculosis* (appendicitis-like symptoms more often than diarrhea).

Q: What are the principal findings in a patient with *Brucella* infection?

A: Patients typically have intermittent fever (classically, there is a fever spike in the afternoons up to 104°F), night sweats, headache, muscle weakness and fatigue, arthralgia (joint pain), and chronic infection lasting as long as 12 months. Patients commonly have enlarged livers and/or spleens. As with tuberculosis and syphilis, it can imitate other diseases. Brucellosis is the most common zoonosis in the world, most notably in developing countries.

Q: IgM antibody specific to West Nile virus was detected in the CSF of a patient. What does this suggest?

A: IgM antibody does not cross the blood–brain barrier. Therefore IgM antibody in CSF suggests CNS infection.

Q: Many anaerobic bacteria can survive in the presence of oxygen until anaerobic conditions develop. Name the two enzymes that these bacteria may possess that allow for this survival.

A: Superoxide dismutase and catalase are two of the most common, but some anaerobes might have other enzymes protecting them from oxygen toxicity.

Q: A 58-year-old male vagrant was admitted into the emergency room with a necrotic wound infection on his foot. The anaerobic organism that was isolated showed a double zone of hemolysis on a blood agar plate. What is the most likely organism (genus and species) and what is the drug of choice?

A: *Clostridium perfringens* and penicillin. Always include anaerobes when dealing with necrotic wound infections. Be aware, however, that anaerobes may not be involved in all necrotic infections.

Q: A patient presents with a wound on his knuckles that he received when hitting his "buddy" on the mouth during a fight at a bar. What bacterium is likely?

A: *Eikenella corrodens* is the most likely cause of a clenched fist injury, which is characterized by a rapid onset of swelling and erythema. Consideration should also be given to *Staphylococcus aureus*, viridans streptococci, *Streptococcus pyogenes*, or other oral flora, so pay attention to additional clues that may be given in the stems of questions. *Eikenella* is part of the HACEK group (*Haemophilus, Actinobacillus, Cardiobacterium, Eikenella,* and *Kingella*) that can be causes of culture-negative endocarditis.

Q: What is a carbuncle and what is the most common infectious cause?

A: A deep abscess similar to a boil (but larger) that interconnects and extends into the subcutaneous tissue. They typically originate as infected hair follicles. Carbuncles are almost always due to *Staphylococcus aureus*.

Q: Name two genera of obligate intracellular bacteria.

A: *Chlamydia* and *Rickettsia* are the two that most think of, but *Chlamydophila, Coxiella* (Q fever), *Ehrlichia,* and *Anaplasma* are others.

Q: What do *Chlamydophila* spp. cause?

A: The genus *Chlamydia* has undergone taxonomic changes. Some species retain the name *Chlamydia* (*Chlamydia trachomatis*), whereas others have been renamed *Chlamydophila* (*Chlamydophila pneumoniae*, pneumonia and bronchitis; *Chlamydophila psittaci*, psittacosis).

Q: What does *Anaplasma* cause?

A: Human granulocytic anaplasmosis (formerly human granulocytic ehrlichiosis)

Q: Name some genera of facultative intracellular bacteria.

A: *Mycobacterium, Salmonella, Shigella, Listeria, Legionella,* and *Brucella* are examples.

Q: What is the most common source of *Bacillus anthracis* infection?

A: Cattle and sheep are classically the most common sources of infection. The most common and most easily treated form is cutaneous anthrax. GI and lung involvement may rapidly lead to death.

Q: What does *Corynebacterium diphtheriae* look like microscopically?

A: Gram-positive rods that resemble Chinese characters

Q: What medium is used in the laboratory to grow *Corynebacterium diphtheriae*?

A: Loeffler and Tellurite media are two common.

Q: What medium is used in the laboratory to grow *Bordetella pertussis*?

A: The organism is plated on Bordet-Gengou agar that is supplemented with sheep blood.

Q: What are the major causes of nongonococcal urethritis?

A: *Chlamydia trachomatis*, mycoplasmas (*Ureaplasma urealyticum, Mycoplasma hominis, Mycoplasma genitalium*), *Trichomonas vaginalis*. Urethritis due to most infectious agents is considered an STD.

Q: What is the drug of choice to treat close contacts and family members exposed to a patient with *Haemophilus influenzae*?

A: Rifampin

Q: What is the mode of action of the fluoroquinolones such as ciprofloxacin?

A: They inhibit DNA synthesis by blocking DNA gyrase activity.

Q: What media are used for the laboratory diagnosis of *Neisseria gonorrhoeae*?

A: Thayer-Martin agar as the selective medium and a nonselective medium such as chocolate agar

Q: An outbreak of food poisoning occurred in August following a family reunion being held at a lake. Fourteen people were affected, most within 48 hours, with symptoms ranging from mild to severe diarrhea, fever, abdominal pain, and occasional vomiting. MacConkey agar yielded colorless colonies of gram-negative bacteria that were motile and produced hydrogen sulfide. In most cases of this food poisoning, what is the treatment?

A: Fluids and electrolytes to prevent dehydration. This is most likely *Salmonella* food poisoning and generally is self-limiting. Antibiotics can result in a carrier state (prolonged shedding) and the development of resistance. Clues in the question: relatively rapid onset compared with most other bacterial food poisonings (but longer than *Staphylococcus aureus*, *Clostridium perfringens*, and *Bacillus cereus*), typical symptoms for salmonellosis, lactose nonfermenter (colorless colonies on MacConkey agar, similar to *Shigella*), motile (*Shigella* is nonmotile), and hydrogen sulfide positive (*Shigella* is negative).

Q: A liquid stool specimen from a patient with diarrhea was submitted to the clinical microbiology laboratory. Gram stain of the stools revealed many gram-negative bacilli, occasional gram-negative curved rods and cocci, as well as gram-positive bacilli and cocci. What can you conclude about the probable cause of the diarrhea?

A: Not much. This is a diarrhea of unknown etiology. Microscopy reveals only normal flora. There are not many bacterial diarrheas in which microscopy is useful.

Q: A 47-year-old man called his physician complaining of vomiting, some abdominal cramping, and mild diarrhea. His temperature was not elevated. Earlier in the day he had attended a wedding reception and had sampled most of the foods (tuna, ham, chicken, and pasta salads, "hot wings," cold meat sandwiches, cheeses, cream-filled pastries, cake, etc.). The observed symptoms are most likely due to what bacterial virulence factor?

A: An enterotoxin produced in one (or more) of the foods by *Staphylococcus aureus*. Clues: typical foods for *S. aureus* food poisoning (salads, cold meat, and pastries), rapid onset, and typical symptoms (especially lack of fever and the prominent vomiting; the toxin is a neurotoxin that affects the vomiting center). Note that an enterotoxic strain of *Bacillus cereus* can also cause food poisoning with similar symptoms.

Q: From the previous question, how should most cases of this food poisoning be treated?

A: Fluids and electrolytes to prevent dehydration. Staphylococcal food poisoning is rapid onset and short duration as is that due to *Bacillus cereus*. In most cases, the specific pathogen is not important.

Q: What was the cause of the "Black Death"?

A: *Yersinia pestis*. Plague is spread by fleas.

Q: Name some bacteria that secrete IgA protease.

A: The most important are *Haemophilus influenzae*, *Streptococcus pneumoniae*, *Neisseria meningitidis*, and *N. gonorrhoeae*.

Q: What is the importance of IgA protease secretion?

A: It aids pathogenic bacteria in colonization of mucous membranes.

Q: The incidence of *Mycoplasma pneumoniae* respiratory disease is highest in what age group?

A: Young adults

Q: Describe the laboratory identification of *Legionella*.

A: Two of the most important things to remember about *Legionella* are (1) they are difficult to see with the Gram stain and (2) they do not grow on typical laboratory media (i.e., blood agar). Therefore special stains such as a silver stain are used for microscopy. Buffered charcoal-yeast extract agar containing L-cysteine and iron is used for growth. In addition, there are a number of serological tests for antigens in the blood or urine. Besides *L. pneumophila*, there are over 50 additional species of *Legionella*, at least 20 of which have been isolated from clinical specimens.

Q: What are the two kinds of respiratory disease due to *Legionella*?

A: Legionnaires' disease (legionellosis), which is characterized by fever, myalgia, cough, pneumonia, and Pontiac fever (a mild "walking" pneumonia).

Q: How is *Legionella* transmitted between people?

A: Trick question! Person-to-person transfer has not been demonstrated.

Q: List the virulence factors of *Legionella*.

A: One of the major virulence factors is the fact that it is a facultative intracellular parasite capable of surviving within alveolar macrophages and monocytes. Growth within these cells protects the bacteria from the humoral immune system. Survival within the macrophages is due in part to the ability of the organism to prevent phagosome–lysosome fusion. Intracellular growth and the production of several additional enzymes result in death of the host cell.

Q: Differentiate between bacterial food poisoning etiologies based on time of onset of symptoms.

A: Rapid onset (less than 18 hours): *Staphylococcus aureus*, *Clostridium perfringens*, and *Bacillus cereus*. Slow onset (more than 24 hours): *Salmonella*, *Shigella*, and *Campylobacter*. These are the major ones you need to know.

Q: With what kind of food is *Salmonella enterica* most commonly associated?

A: All serotypes of *Salmonella* that cause gastroenteritis are commonly associated with poultry but can be found in other foods such as hamburger. They can also be found in fruits, vegetables, and other foods that become contaminated.

Q: Cold enrichment can be used for the isolation of what organism?

A: *Listeria monocytogenes*. Cold enrichment has been used to enrich for this organism in a mixture of other bacteria that do not grow at low temperatures. Infection due to *Listeria monocytogenes*, which is a small gram-positive coccobacillus, has frequently been associated with refrigerated dairy products but has also been associated with other foods (various meats, unwashed vegetables, etc.).

Q: What is the primary reservoir for *Yersinia pestis*?

A: Wild rodents and their infected fleas. Humans normally are infected via a flea bite. The major endemic area in the United States is in the southwest.

Q: How is *Y. pestis* spread from person-to person?

A: Person-to-person spread is by aerosol droplets from patients who have pneumonic plague. This form, which is not found in the United States, has a shorter incubation time and higher mortality rate than does the bubonic plague.

Q: Can the bubonic form of *Y. pestis* infection progress to the pneumonic form?

A: Yes. If hematogenous spread results in lung involvement, then it can develop into the pneumonic form.

Q: How do neonatal *Listeria monocytogenes* infections occur?

A: In utero by transplacental infection or during (or just after) delivery. In utero infection has an earlier onset, and it is much more devastating than is infection occurring after delivery.

Q: What is the composition of the capsule of *Streptococcus pneumoniae*?

A: Polysaccharide. Most bacterial capsules are polysaccharide. The major exception is the polyglutamic acid capsule of *Bacillus anthracis*.

Q: What causes amebic dysentery and how is it treated?

A: *Entamoeba histolytica*, which is commonly treated with metronidazole or tinidazole

Q: How is *E. histolytica* spread and what is the infectious form?

A: It is most commonly spread by food or water contaminated with cysts. Ingested trophozoites (active motile form) are not infectious.

Q: What is the mechanism of staphylococcal toxic shock syndrome toxin-1?

A: It is a superantigen that binds to the MHC class II antigen on an antigen-presenting cell and the variable portion of the beta chain on the T-cell receptor, causing the nonspecific activation of T cells and the release of large amounts of cytokines such as IL-1, IL-2, and TNF-alpha.

Q: What is the most common STD in sexually active teenage males?

A: Nongonococcal urethritis due to *Chlamydia trachomatis*. *C. trachomatis* is the most common sexually transmitted pathogen in developed countries.

Q: A 1-day-old neonate who developed respiratory distress and seizures was diagnosed with pneumonia, bacteremia, and meningitis. The gram-positive coccus that was isolated from blood and CSF is most likely what?

A: *Streptococcus agalactiae*. This is an example of early onset disease.

Q: Which valve does subacute bacterial endocarditis most commonly affect?

A: Subacute bacterial endocarditis usually involves the left side of the heart, specifically the mitral valve. The aortic valve is the second most commonly involved valve. Rheumatic fever is the most probable underlying cause of valvular damage associated with subacute bacterial endocarditis. Mitral stenosis is a common predisposing factor. Drug addicts tend to develop right-sided subacute bacterial endocarditis, usually involving the tricuspid valve.

Q: What is the cause of Q fever?
A: *Coxiella burnetii*

Q: How is *Coxiella burnetii* classified?
A: It is a rickettsia.

Q: How is *C. burnetii* spread?
A: Via dust particles. *C. burnetii* causes Q fever, which is a zoonotic disease. In infected mammals the organism has a predilection for the placenta. After giving birth, dried placentas on the ground serve as a source of infection. Besides mammals, other animals such as birds can be infected. Infective particles can survive in soil for years.

Q: How does Q fever present?
A: Although there can be severe manifestations, which you do not need to learn about at this time, it normally presents as a mild respiratory disease similar to atypical (or "walking") pneumonia. As with typical rickettsial diseases, Q fever usually begins with fever, chills, and headache.

Q: Describe the spread of the rash of Rocky Mountain spotted fever.
A: It typically appears on the hands and feet and rapidly spreads to the trunk. It is caused by *Rickettsia rickettsii*.

Q: What is the animal reservoir of *Salmonella typhi*?
A: Trick question! *S. typhi* causes typhoid fever and is strictly a human pathogen. It is not found in animals. All other serovars of *Salmonella enterica* (*S. typhimurium*, *S. enteritidis*, etc.) that cause gastroenteritis are found in the intestinal tracts of a large variety of animals, including domestic livestock, poultry, reptiles, and amphibians.

Q: What species of *Shigella* is the most common cause of shigellosis in the United States?
A: *Shigella sonnei*. Other species include *S. dysenteriae* (causes the most severe disease), *S. flexneri* (more common in developing countries), and *S. boydii*.

Q: For what kinds of infections can metronidazole be used?
A: Infections due to anaerobic bacteria as well as some protozoa (*Giardia*, *Entamoeba* and *Trichomonas*)

Q: What type of cell in the intestinal tract does *Shigella* typically invade?
A: *Shigella* typically invades the M cells of the Peyer patches where it grows in the cytoplasm. Other intestinal pathogens including *Salmonella*, *Yersinia*, other bacteria, and some viruses invade M cells.

Q: Describe the clinical presentation of shigellosis.
A: Abdominal cramps are accompanied with liquid stools containing pus, mucus, and blood. Typically, it is a self-limited disease requiring only fluids and electrolytes. Serious cases may require an antimicrobial such as a fluoroquinolone.

Q: What kind of organism causes scabies?
A: It is a mite (*Sarcoptes scabiei*). Scabies is characterized by intense itching (delayed hypersensitivity).

Q: What yeast has broad-based buds?
A: *Blastomyces dermatitidis*

Q: What is the composition of the two primary vaccines for *Streptococcus pneumoniae*?
A: The standard pneumococcal polysaccharide vaccine (PPV or PPV23) is composed of 23 serotypes of the capsular polysaccharide. For children under 2, the heptavalent pneumococcal conjugate vaccine (PCV7) is used (seven capsular serotypes).

Q: At what age should the pneumococcal polysaccharide vaccine (PPV23) normally be given?
A: Generally, at age 65 or greater. Others who should receive the pneumococcal vaccine include those with chronic cardiorespiratory disease, cirrhosis, alcoholism, chronic renal disease, diabetes mellitus, splenic disorders, sickle-cell disease, HIV and others who may be immunocompromised.

Q: Who and at what age should the PCV7 be given?
A: All children should receive four doses of PCV at ages 2, 4, 6, and 12–15 months of age.

Q: In what disease is the detection of Donovan bodies diagnostic?

A: Granuloma inguinale, also called donovanosis. This STD is caused by the bacterium *Klebsiella granulomatis* (previously *Calymmatobacterium granulomatis*). Donovan bodies are the bipolar staining bacteria within histiocytes and are diagnostic for this disease. You probably only need to know that this disease is an STD and *maybe* be able to recognize the name of the organism. It is unlikely that you will need to know about Donovan bodies for a standardized exam, but just in case, I have included an explanation. It is more likely that medical students will get this question from a resident in their third- or fourth-year rotation.

Q: Describe the signs and symptoms of Guillain-Barré syndrome.

A: Guillain-Barré disease classically presents with symmetrical weakness in the legs, which ascends to include the arms or trunk. Distal weakness is usually greater than proximal. The onset of the disease rarely involves the cranial nerves. The disease has multiple etiologies including noninfectious (autoimmune, malignancies, surgery, drugs) and infectious, especially infection with *Campylobacter jejuni* (about half of all cases in the United States), but also *Mycoplasma pneumoniae*, CMV, EBV, and HIV. Some cases have been associated with vaccinations (e.g., influenza and rabies).

Q: Which are more common to gram-negative bacteria: exotoxins or endotoxins?

A: Virtually all gram-negative bacteria contain an endotoxin. Its structure may vary among different species (e.g., the endotoxin of *Neisseria* contains lipo-oligosaccharide rather that LPS). In addition, many gram-negative bacteria, such as *Shigella*, *Vibrio*, *Pseudomonas*, and some strains of *E. coli*, also secrete exotoxins. Not all endotoxins have the same biological activity in the body. Some are important virulence factors and others contribute little to the disease.

Q: What does *Borrelia recurrentis* cause?

A: Relapsing fever

Q: How is *Borrelia recurrentis* transmitted?

A: By lice. Other species of *Borrelia* can cause tick-borne relapsing fever. Neither form is common in the United States.

Q: How is leptospirosis spread?

A: The causative organism, *Leptospira interrogans*, is found in many types of mammals, including pet dogs and urban rats. Infection can occur after contact with animal urine or contact with food or water (either drinking or swimming) contaminated with animal urine.

Q: A 37-year-old woman presented with abdominal pain, weight loss, vaginal discharge, and fever. An intrauterine device she had been using for birth control for three years was removed and sent to the laboratory for culture. The organism that was isolated from the surface of the intrauterine device was an obligately anaerobic gram-positive branching bacillus. What is the most probable organism (genus and species), and what is the drug of choice?

A: *Actinomyces israelii* is the bug and penicillin is the drug. Although other bacteria may be involved, you should be familiar with the association of *A. israelii* infection and intrauterine device usage.

Q: A sexually active teenage girl presents with fever, rash, and pain in her knees and elbows. She just completed her menses. Physical examination reveals fever; a swollen, red, hot knee; and erythematous macules of her hands and soles of her feet. What is the most likely diagnosis?

A: Disseminated gonococcal infection

Q: What is the incubation time of *Salmonella* gastroenteritis?

A: Approximately 1–2 days after consumption of the contaminated food compared with GI upset caused by *Staphylococcus*, which has a much shorter incubation period of approximately six hours.

Q: Which species of *Staphylococcus* is most commonly associated with catheter infections?

A: *Staphylococcus epidermidis*

Q: What bacterial/viral diseases can cause Bell (facial) palsy?

A: Both Lyme disease and herpes zoster (shingles) can result in Bell palsy. It can also occur in pregnant women, those with diabetes, persons with influenza or the common cold, and in people who are immunocompromised.

Q: What virulence factor allows *S. epidermidis* to adhere to artificial materials such as shunts or catheters?

A: The polysaccharide slime layer (capsule) of *S. epidermidis* sticks tenaciously to artificial surfaces where it grows as a biofilm.

Q: Which species of *Staphylococcus* is commonly associated with UTIs primarily in young women?

    A: *Staphylococcus saprophyticus* causes 10–20% of UTIs in young females.

Q: What type of infection can be characterized as a localized infection of the superficial skin layers, presenting as pus-filled pustules that turn crusty when they break, and is commonly seen in infants and young children?

    A: Impetigo (pyoderma)

Q: What bacteria most often cause impetigo?

    A: The two most common causes are *Staphylococcus aureus* (80%), usually presenting as bullous impetigo, and *Streptococcus pyogenes* (20%), usually presenting as a crusty impetigo.

Q: What rash is typically due to a pyrogenic toxin of *Streptococcus pyogenes*?

    A: Scarlet fever

Q: What streptococci typically cause neonatal sepsis?

    A: *Streptococcus agalactiae* (group B)

Q: Describe the differences between alpha-, beta-, and gamma-hemolysis.

    A: Alpha-hemolysis is incomplete lysis of red blood cells resulting in a green pigment surrounding the colony on blood agar plates. Beta-hemolysis is total or complete lysis of red blood cells resulting in a clear area around the colony on a blood agar plate. In gamma-hemolysis, there is no hemolysis on a blood agar plate. Bacteria in this last group are more appropriately called nonhemolytic.

Q: What is the hemolytic pattern of *Streptococcus pyogenes*?

    A: It is beta-hemolytic (complete or clear).

Q: What are the distinguishing features of the viridans group of streptococci?

    A: They are part of the normal oral flora, are not inhibited by optochin, and are alpha-hemolytic.

Q: What diseases are associated with *Streptococcus pneumoniae* and what are its distinguishing features?

    A: *Streptococcus pneumoniae* is a gram-positive diplococcus that causes pneumonia (especially in children and the elderly, but all ages are susceptible), otitis media in children, acute rhinosinusitis, and meningitis. It has a large antiphagocytic polysaccharide capsule, is sensitive to optochin, lysed by bile, and is alpha-hemolytic.

Q: What bacteria cause bacteremia?

    A: Almost all. With questions concerning bacteremia, read the question carefully, looking for clues that guide you to the most probable organism. Bacteremia is only the presence of bacteria in the blood and does not necessarily indicate serious disease. Septicemia implies growth of the bacteria in the blood and is frequently serious. The terms "bacteremia" and "septicemia" are often, and incorrectly, used interchangeably.

Q: What infections are commonly caused by *Streptococcus agalactiae*?

    A: Neonatal meningitis and sepsis. It is commonly referred to as group B strep.

Q: What diseases are commonly associated with *Haemophilus influenzae*?

    A: Otitis media, conjunctivitis, pneumonia, arthritis, osteomyelitis, meningitis, cellulitis, and epiglottitis. Those at highest risk are nonimmunized or inadequately immunized children and the elderly with a weakened immune response. Because of the use of the Hib vaccine, the incidence of disease in children has been decreasing.

Q: What are the distinguishing structural and biochemical features of *Streptococcus pyogenes*?

    A: It possesses M protein in its cell wall, is bacitracin sensitive, catalase negative (as are all streptococci), excretes pyrogenic exotoxins as well as streptolysins S and O, and is beta-hemolytic.

Q: What diseases are commonly associated with *S. pyogenes*?

    A: Impetigo, pharyngitis, scarlet fever, rheumatic fever, cellulitis, erysipelas, endocarditis, acute glomerulonephritis, necrotizing fasciitis, and toxic shock syndrome

Q: Name four common parasitic diseases of humans in the United States.

    A: The four most common are giardiasis, pinworm infection, trichomoniasis, and toxoplasmosis.

Q: Reduviid bugs are vectors for what disease?

    A: Chagas disease

Q: Where is Chagas disease endemic?

    A: It is most common in Central and South America and is rare in the United States.

Q: What is the cause of river blindness?

A: *Onchocerca volvulus*, but if you know just *Onchocerca* that should be enough. The disease is also called onchocerciasis.

Q: A 27-year-old burn wound patient with a urinary catheter developed a UTI. A gram-negative urease positive organism was isolated on MacConkey agar. An additional unique characteristic of this organism on a blood agar plate is that it swarmed. This bacterium is most likely what?

A: *Proteus*. There are several different species of *Proteus*, including *P. mirabilis* and *P. vulgatus*. You will not be expected to be able to differentiate between the two. They most commonly cause UTIs.

Q: What role does the urease of *Proteus* play in UTIs?

A: Urease splits urea into ammonia and carbon dioxide. The ammonia raises the pH of the urine to alkaline, causing the precipitation of divalent cations, especially calcium and magnesium, resulting in stone formation (urinary calculi/struvite stones).

Q: What common opportunistic pathogen produces a bluish-green pigment?

A: *Pseudomonas aeruginosa*. This is one of the most important identifying characteristics of this organism. The bluish-green color is actually due to two pigments, the blue pyocyanin and the yellow fluorescein, which fluoresces when exposed to UV light.

Q: A 73-year-old woman visited her physician for a routine medical evaluation. Her vitals were within normal ranges for her medical condition. She reported that she felt well. A routine quantitative urine culture resulted in 1.7×10^5 cfu/mL. Based on this information, should she be treated for a UTI?

A: No. Although quantitative urine culture results greater than 100,000 cfu/mL in an asymptomatic individual indicate a UTI, not all such patients need to be treated, including geriatric patients. Asymptomatic UTIs occur most commonly in the elderly and patients with a urinary catheter. Pregnant females with an asymptomatic UTI should be treated because of possible complications such as preterm birth or preeclampsia.

Q: Several children at a local day care in Santa Ana, California were sent home because of diarrhea. All had abdominal cramps and mild to moderate fever. Vomiting was not a common feature. One child had mucus and blood in his stools. Laboratory culture yielded nonlactose fermenting, gram-negative, facultatively anaerobic rods that were nonmotile and hydrogen sulfide negative. What is the most likely etiological agent?

A: *Shigella sonnei*. Nonlactose fermenting bacilli indicate either *Salmonella* or *Shigella*. Of these two, only *Shigella* is nonmotile as well as hydrogen sulfide negative. *Shigella sonnei* is most common in the United States.

Q: A gram-positive, coagulase-positive coccus isolated from ham was suspected of being the cause of an outbreak of food poisoning at an Easter buffet at a local restaurant. What is the major virulence factor of this organism in this disease?

A: The organism secretes an enterotoxin in the food. It is most probably a case of staphylococcal food poisoning.

Q: In an outbreak of food poisoning at a fraternity house, symptoms occurred within a day for most of the students. No obvious pathogens were isolated from stool specimens with routine culture, but bacteria were eventually isolated from some chicken gravy. The bacteria showed a characteristic double zone of hemolysis and resembled boxcars on Gram stain. What is the major virulence factor of this organism in this disease?

A: An enterotoxin is secreted during sporulation in the gut. This is a case of *Clostridium perfringens* food poisoning.

Q: What is type I hypersensitivity? Give some examples.

A: This is immediate hypersensitivity (IgE mediated), also known as the allergic reaction. IgE that is bound to Fc receptors on tissue mast cells and circulating basophils serves as receptors for the antigen (i.e., allergen). The antigen that is bound by two adjacent IgE molecules results in cross-linked (via the antigen) IgE on the surface of the mast cells and basophils. This causes degranulation of the mast cells and basophils and the release of histamine and other primary and secondary mediators. Examples of type I hypersensitivity include drug (e.g., penicillin), food (e.g., peanuts), and some parasite (e.g., helminths) allergies, as well as anaphylactic shock, allergic asthma, hay fever, and urticaria.

Q: What is type II hypersensitivity? Give some examples.

A: This is also called cytotoxic hypersensitivity and involves the binding of either IgG or IgM to cell surface associated antigens, followed by complement activation and resulting in cell lysis. Type II hypersensitivity can also result in phagocytosis by macrophages and neutrophils. Examples include Goodpasture syndrome, Rh and ABO blood group incompatibility reactions, and myasthenia gravis.

Q: What is type III hypersensitivity? Give some examples.

A: This is immune complex mediated (complexes of antigen and IgG or IgM) and results in tissue damage due to complement activation. Polymorphonuclear neutrophilic leukocytes are commonly attracted to the deposition site and are involved in the concomitant inflammatory response. Classic examples are the Arthus reaction and serum sickness. Other examples include poststreptococcal glomerulonephritis, systemic lupus erythematosus, and farmer's lung.

Q: What is type IV hypersensitivity? Give some examples.

A: There are two kinds: delayed-type hypersensitivity (also called cell-mediated hypersensitivity) and T-cell mediated cytotoxicity. Delayed-type hypersensitivity involves sensitized CD4+ T lymphocytes. Foreign antigens on antigen-presenting cells make contact with CD4+ T cells, resulting in T-cell activation, the release of various cytokines (e.g., IL-2, TNF, and interferon-gamma), and the activation of macrophages. Examples include the tuberculin skin test, contact dermatitis, and tuberculosis. T-cell mediated cytotoxicity involves CD8+ T cells that kill cells that have certain antigens on their surface. Examples include the destruction of virus-infected cells (class I molecules complexed to viral peptides on cell surfaces) and graft (transplant) rejection (MHC molecules on cell surface).

Q: What is the paratope of an antibody molecule?

A: It is the site on the antibody that binds to a specific site (epitope) on the antigen and is composed of the variable domains of the heavy and light chains.

Q: What are cytokines?

A: Small peptides that serve to regulate the immune system. Many cell types, especially macrophages and lymphocytes, produce them. They include the interleukins (IL-1, IL-2, etc., TNF, interferons, colony-stimulating factors, etc.). Cytokines may have other non-immune system regulatory functions.

Q: What kinds of cells secrete antibodies?

A: B cells and plasma cells (think of plasma cells as mature B cells)

Q: What type of antibody is found in mucus?

A: IgA in its dimeric form (secretory IgA). IgA is also the predominant antibody in other secretions such as milk and tears, but it is also common in its monomeric form in serum.

Q: What holds the two chains together to make a dimer of IgA?

A: The J (joining) chain

Q: What is the primary immunoglobulin in serum?

A: IgG (about 85%)

Q: What is a hapten?

A: A small molecule that by itself cannot induce an immune response

Q: What is the function of CD4+ T cells?

A: They are helper T lymphocytes (helper T cells or T_H cells) and initiate a specific immune response when the T-cell receptor on the surface of the T_H cell comes in contact with a processed antigen that is in the groove of the MHC class II molecule on the surface on an antigen presenting cell. T_H cells can be further divided into two subsets as they mature (after contact with an antigen-presenting cell), depending on a number of factors (type of antigen-presenting cell, antigen concentration, cytokine activation, etc.): T_H1 and T_H2 cells. T_H1 cells are involved in cytotoxic T-cell induction, and T_H2 cells are involved in antibody production by B cells.

Q: What is the function of cytotoxic T cells?

A: Cytotoxic T cells (they are also called CD8+ cytotoxic T cells or Tc cells) kill cells such as tumor cells or virus-infected cells. They also function in killing cells in graft rejection. T_H1 cells are involved (via cytokine secretion) in the activation of Tc cells.

Q: What is urticaria?

A: Hives (type I hypersensitivity)

Q: On first contact with the epitope of an antigen, only a few B cells are able to recognize the epitope. The immune response is weak. Four years later when encountering the same antigen, there is a much stronger and more rapid response to that specific antigen. What cells are responsible for this response?

A: Memory cells. Both B and T cells can be memory cells.

Q: Antigen-presenting cells present the processed antigen to what type of cells?

A: Antigen-presenting cells present exogenous antigens (e.g., bacteria components) to helper T cells in the groove of MHC class II molecules. Endogenous antigens (e.g., from tumors or viruses) are presented to cytotoxic T cells in the groove of MHC class I molecules.

Q: What antimicrobials are commonly used to control epidemics of meningococcal meningitis?

A: Rifampicin, ciprofloxacin, and ceftriaxone are some of the more common

Q: What is the presumptive diagnosis of gram-positive, lancet-shaped, optochin-sensitive diplococci that form alpha-hemolytic colonies on blood agar plates?

A: *Streptococcus pneumoniae*

Q: LPS is the same as what?

A: Endotoxin

Q: What structure of bacteria gives shape to individual cells?

A: Peptidoglycan (murein)

Q: In transformation as it occurs in bacteria, what is the source of the DNA?

A: The DNA can potentially be from any kind of cell, prokaryotic or eukaryotic. Transformation is the uptake of soluble DNA. After the DNA gets into the cell, it can integrate into the bacterial chromosome.

Q: List some antimicrobials that inhibit folate synthesis.

A: The primary agents are the sulfonamides and trimethoprim, but there are drugs such as dapsone (a sulfone) and para-aminosalicylic acid that have this mechanism. The sulfonamides, sulfones, and para-aminosalicylic acid are analogs of para-aminobenzoic acid. Trimethoprim is also a metabolic analog. It inhibits tetrahydrofolic acid production in bacteria by binding to and inhibiting the enzyme dihydrofolate reductase.

Q: Describe halophilic bacteria and give some examples.

A: Halophiles require increased salt concentrations for growth. Examples of human pathogens include *Vibrio parahaemolyticus* (primarily seafood associated gastroenteritis but occasionally wound infections) and *Vibrio vulnificus* (seawater-associated sepsis and gastroenteritis). *Staphylococcus aureus* and *Enterococcus* are facultative halophiles and therefore are capable of growing at elevated salt concentrations.

Q: How are bacteria that grow best at moderate temperatures (e.g., 25–37°C) classified?

A: They are mesophiles. Most bacterial pathogens are mesophiles.

Q: What is the source of human infection with *Listeria monocytogenes*?

A: Most cases are food-borne, often associated with contaminated dairy products and meats. Neonates can be infected transplacentally as well as during or after birth.

Q: Infections with what bacterium can result in a pseudomembrane that can lead to suffocation?

A: *Corynebacterium diphtheriae*

Q: *Borrelia burgdorferi* is the cause of what disease?

A: Lyme disease

Q: The causative agent of typhoid fever is what?

A: *Salmonella typhi*

Q: Order the following from the least amount of energy (in the form of ATP) to the most amount of energy that bacteria can generate: anaerobic respiration, fermentation, aerobic respiration.

A: Fermentation (usually net gain of 2 ATPs), anaerobic respiration (less then 38 ATPs, depending on terminal electron acceptor), aerobic respiration (38 ATPs). You do not need to memorize the number of ATPs generated by each process. It can be variable, depending on the species and other factors.

Q: What gram-negative organism is associated with gastritis and peptic ulcer disease?

A: *Helicobacter pylori*

Q: Name two genera of pathogens that require a microaerophilic atmosphere for growth.

A: *Campylobacter* and *Helicobacter*. *Borrelia* is also, but you do not need to know that.

Q: What gram-negative, rod-shaped bacterium is aerobic, motile, oxidase positive, oxidizes glucose, has a prominent capsule, and produces a blue-green pigment?

A: *Pseudomonas aeruginosa*

Q: What are the primary pathogens in acute bacterial rhinosinusitis?

A: *Streptococcus pneumoniae* and *Haemophilus influenzae*. *Moraxella catarrhalis*, *Staphylococcus pyogenes*, *Staphylococcus aureus*, and various anaerobes are less frequent. Most cases of acute rhinosinusitis are viral. Chronic sinusitis can have both infectious and noninfectious causes. In acute sinusitis it is important to demonstrate a bacterial etiology before antibiotics are prescribed. Most patients with acute rhinosinusitis recover without antibiotic therapy, but it should be considered in patients with prolonged or more severe symptoms.

Q: The two growth factors hematin (X) and NAD (V) are required for the growth of what organism?

A: *Haemophilus influenzae*

Q: *Clostridium difficile* is commonly associated with what disease?

A: Antibiotic-associated pseudomembranous colitis is what usually comes to mind, but it is a more common cause of antibiotic-associated diarrhea (usually precedes colitis).

Q: What are the granulomatous lesions of tertiary syphilis called?

A: Gummas. They are not infectious.

Q: Gummas of syphilis in the perianal region are called what?

A: Condyloma latum (genital warts)

Q: Gumma-like lesions of papillomavirus in the perianal region are called what?

A: Condyloma acuminatum (genital warts)

Q: What bacterial infection can lead to gastric cancer?

A: *Helicobacter pylori* gastritis

Q: After infection by a bacteriophage, a strain of *Clostridium botulinum* was able to synthesize and secrete an exotoxin. This is most likely an example of what genetic event?

A: Lysogenic conversion

Q: What clinical and laboratory findings are necessary to make the diagnosis of bacterial vaginosis due to *Gardnerella vaginalis*?

A: A whitish adherent vaginal discharge; vaginal pH > 4.5 (normal is pH 3.8–4.5); amine-like (fishy) odor generated after adding 10% potassium hydroxide (KOH) to a sample of the vaginal discharge (whiff test); demonstration by microscopy that more than 20% of vaginal epithelial cells are covered with gram-negative coccobacilli (clue cells)

Q: A yeast that was isolated from a patient gave a positive germ tube test. What is the most likely species of the yeast?

A: *Candida albicans*

Q: Encapsulated yeast cells seen microscopically in an India ink preparation of CSF indicate an infection with what?

A: *Cryptococcus neoformans*

Q: Anatomically, where is the initial infection with *C. neoformans* normally found?

A: Lungs. Cryptococcal meningitis can occur after hematogenous spread.

Q: How is *C. neoformans* infection acquired?

A: Inhalation of dust contaminated with bird (especially pigeon) droppings

Q: What name is given to systemic tuberculosis that results in many tubercles in an organ?

A: Miliary tuberculosis

Q: What is the hemolytic pattern of *Streptococcus pyogenes* on a blood agar plate?

A: It is beta-hemolytic.

Q: What is the name given to the infectious form of chlamydia?

A: Elementary body

Q: Most cases of primary atypical pneumonia due to *Mycoplasma pneumoniae* are seen in what age group?

A: Five to 25 years. Questions concerning *Mycoplasma pneumoniae* often include a patient who is an older teenager or who is in their early 20s.

Q: What causes the tissue damage in immune-complex mediated (type III hypersensitivity) glomerulonephritis?

A: Initiation of the classic complement cascade. The antigen–antibody complexes that form in some diseases are trapped in the capillaries of the kidneys where there is an initiation of the complement cascade.

Q: Name some infectious diseases and their etiological agents that are tick-borne.

A: Colorado tick fever (due to an orbivirus in the family Reoviridae), Rocky Mountain spotted fever (*Rickettsia rickettsii*), relapsing fever (*Borrelia recurrentis*), Lyme disease (*Borrelia burgdorferi*), tularemia (*Francisella tularensis*; there are other ways of acquiring tularemia), Powassan encephalitis (Powassan virus, a flavivirus), ehrlichiosis (human monocytic ehrlichiosis; *Ehrlichia*), anaplasmosis (human granulocytic anaplasmosis [previously human granulocytic ehrlichiosis]; *Anaplasma*), and babesiosis (*Babesia*) are some.

Q: What is the composition of viral capsids?

A: Protein. In enveloped viruses the shell is composed of the nucleic acid plus the capsid and it is called the nucleocapsid.

Q: List some of the major causes of viral diarrhea.

A: Norovirus (older children and adults) and rotavirus (children less than 2) are most common, but others such as adenoviruses, astroviruses, and sapoviruses can cause diarrhea.

Q: What other types of infections do adenoviruses cause?

A: They are important in upper (primarily) and lower respiratory tract infections, especially in children (e.g., the common cold, pneumonia, croup, bronchitis, etc.). Depending on the serotype, they can also cause conjunctivitis, keratoconjunctivitis, and hemorrhagic cystitis. Adenoviruses are a major cause (second behind rotaviruses) of infantile diarrhea.

Q: How is CMV classified?

A: It is a herpesvirus.

Q: What are some CMV disease associations?

A: The most severe infections are acquired in utero and can result in severe CNS manifestations including mental retardation. Other manifestations include premature birth, hepatosplenomegaly, and jaundice. Most human infections are asymptomatic. Severe disease can also occur in AIDS and others who are immunocompromised.

Q: What are flaviviruses and what are some associated infections?

A: They are enveloped single-stranded RNA viruses. Some of the more important diseases include dengue fever, yellow fever, and St. Louis encephalitis.

Q: How are the infections in the previous question transmitted?

A: Mosquitoes

Q: Are tumor viruses usually DNA or RNA viruses?

A: Most are DNA viruses.

Q: Only one group of DNA viruses does not cause tumors in animals or transformation (analogous to neoplasia) of tissue culture cells. What virus group?

A: Parvoviruses

Q: What RNA viruses are tumor viruses?

A: Only one group, the retroviruses

Q: Name an oncogenic retrovirus.

A: Human adult T-cell leukemia virus type I (HTLV-I). It infects T helper cells, leading to impairment of the immune system.

Q: List some neoplasm-associated DNA viruses and the corresponding neoplasm.

A: EBV: Burkitt lymphoma and nasopharyngeal carcinoma; hepatitis B virus: hepatocellular carcinoma; human papillomavirus (especially HPV-16, HPV-18, and HPV-31): cervical neoplasia; HHV-8: Kaposi sarcoma. HHV-8 is also called Kaposi sarcoma–associated herpesvirus. There are other DNA viruses that can either cause tumors in animals or transformation of tissue culture cells, but their association with human neoplasia has not been demonstrated.

Q: What is the classification of the viruses in the previous question?

A: EBV: herpesvirus; hepatitis B virus: hepadnavirus; HPV: papillomavirus; HHV-8: herpesvirus

Q: What is the primary target of HIV-1?

    A: CD4+ T helper cells

Q: Describe the major structural differences between the genomes of the family Orthomyxoviridae and the Paramyxoviridae.

    A: Both are single-stranded RNA viruses, but the orthomyxoviruses have a segmented genome (eight segments) and the paramyxoviruses have only one.

Q: Name some important orthomyxoviruses and disease associations.

    A: There is only one virus in the group, influenza, but three types (A, B, and C). Only types A and B are important in human disease.

Q: Describe the important structural components of the influenza virus.

    A: It is an enveloped virus with two different types of surface spikes, a hemagglutinin (HA), and a neuraminidase (NA). As stated previously, it is a single-stranded RNA virus that has eight segments of RNA.

Q: What are the functions of HA and NA of influenza virus?

    A: The HA functions in attachment of the virion to specific receptors on the cell surface and further augments virus entry into the cell. It can also agglutinate red blood cells from humans and several animal species. The NA functions late in the viral replication process and helps in the release of mature virions from infected cells.

Q: What are the receptors to which the HA binds?

    A: Sialic acid

Q: Explain antigenic shift and drift as they occur in the influenza virus.

    A: Antigenic shift is due primarily to the reassortment of the individual segments of the RNA genome in a host that is infected with two different strains of virus. For example, a pig infected with both H_1N_1 and H_5N_5 can give rise to H_5N_1, the bird flu virus. Shifts are capable of leading to pandemics. Antigenic drifts are due to minor changes from point mutations that affect only one or two amino acids in HA or NA (or both). This can result in smaller more limited epidemics.

Q: Name some important viruses in the family Paramyxoviridae and their disease associations.

    A: Parainfluenza virus (four important serotypes [1–4]): upper and lower respiratory tract infections (cold-like symptoms, pneumonia, croup, bronchitis, etc.); measles virus: measles (rubeola); mumps virus: mumps; respiratory syncytial virus (RSV): most common respiratory pathogen of infants and young children, mild upper respiratory tract infections in older children and adults

Q: Describe the clinical findings in measles.

    A: The incubation period after exposure is about 11 days, followed by cold-like symptoms of fever, cough, coryza (runny nose), conjunctivitis, and photophobia. Within about two days, Koplik spots appear on the buccal surface (across from the molars). Within another day, the maculopapular rash begins on the head, moving down to the trunk and the extremities. After several days, the rash turns brownish and begins to fade in the order it appeared.

Q: How does the rash of measles (rubeola) differ from that of rubella (German measles)?

    A: Both are quite similar and can be difficult to differentiate. Both are maculopapular rashes that begin on the head and move toward the trunk. The rash of measles is a darker red than is the rubella rash. After several days, the rash of measles turns brownish.

Q: What are the complications of measles?

    A: Encephalitis (1 : 1000 cases), subacute sclerosing panencephalitis (very rare), secondary bacterial infections (pneumonia, otitis media)

Q: How is measles transmitted?

    A: Air-borne respiratory droplets or direct contact with respiratory secretions

Q: How is the measles virus classified?

    A: It is a paramyxovirus.

Q: List the major characteristics of paramyxoviruses.

    A: They are enveloped viruses containing a single piece of single-stranded RNA and a helical nucleocapsid.

Q: Describe the clinical manifestations of rubella as might be expected in a child.

A: Acute onset of maculopapular rash (beginning on head, moving toward trunk) with fever, usually accompanied by lymphadenopathy. The mild rash normally lasts three days.

Q: Is rubella infection during pregnancy cause for concern?

A: Yes. It can result in congenital infection. The specific effects on the fetus depend on the stage of development at the time of infection. Infection in the first trimester is most serious and can result in severe defects (mental retardation, congenital heart disease, deafness, cataracts, etc.).

Q: What are Koplik spots? Describe them.

A: Small white spots ("grains of salt") surrounded by a red halo. They typically occur on the buccal mucosa across from the molars about 11 days after infection but before the cutaneous rash. Their appearance is pathognomonic for measles.

Q: What is the most probable etiological agent in an infant who initially presents in February with a low-grade fever and vomiting followed by a watery diarrhea?

A: Rotavirus is the most common cause of gastroenteritis of infants and young children. Most cases occur in the winter in the United States. Low-grade fever is in the range of 99–100°F. Other viruses causing gastroenteritis in infants include adenovirus, astrovirus, and sapovirus. Norovirus (a Calicivirus) also produces infections that cause acute diarrhea and vomiting, but more commonly in older children and adults.

Q: What is the treatment for rotavirus diarrhea?

A: Fluid and electrolyte replacement. For prevention, several live oral vaccines are available.

Q: How is rotavirus infection spread?

A: Fecal–oral

Q: What diseases are commonly associated with bacteria in the viridans group?

A: Dental caries and subacute bacterial endocarditis

Q: Cells that microscopically resemble a "tennis racket" are most likely what bacteria?

A: *Clostridium tetani*. These bacteria have a terminal spore. There are other clostridia that have terminal spores besides *C. tetani*, but you typically do not need to know about how they look microscopically. It should be noted that tetanus is not diagnosed based on the recovery of the organisms.

Q: A widened mediastinum is an important diagnostic sign in what disease caused by a Category A bioterrorism agent?

A: Pulmonary anthrax. A widened mediastinum is not unique to anthrax. It can have other infectious (e.g., acute descending necrotizing mediastinitis due to oral bacteria) and noninfectious (e.g., aortic dissection, lymphoma, others) etiologies.

Q: What causes viral hepatitis?

A: Hepatitis A virus, hepatitis B virus, hepatitis C virus, hepatitis D virus, hepatitis E virus, CMV, EBV, and herpes simplex virus

Q: Describe hepatitis A virus and its epidemiology.

A: Hepatitis A virus is a picornavirus (small RNA virus) that is nonenveloped and contains one segment of single-stranded RNA. Transmission is via the fecal–oral route.

Q: How do you test for viral hepatitis due to hepatitis A?

A: Primarily by detection of specific IgM using serological tests such as enzyme-linked immunosorbent assay or radioimmunoassay

Q: What three genes are present in all retroviruses and what is the function of each?

A: The *gag* gene codes for capsid and core structural proteins; the *pol* gene codes for several essential enzymes, including reverse transcriptase, protease, and integrase; and the *env* gene codes for envelope glycoproteins. These three genes are flanked by long terminal repeats.

Q: What are some causes of aseptic meningitis?

A: HIV, herpes simplex virus, *Leptospira interrogans*, enteroviruses (e.g., Coxsackievirus, echoviruses, enteroviruses, and poliovirus), and arthropod-borne viruses (e.g., Western equine encephalitis virus)

Q: What are Councilman bodies?

A: Liver lesions (cell necrosis due to apoptosis) seen with some viruses, including hepatitis B virus and Dengue virus

Q: What infectious agents cause traveler's diarrhea?

A: Norovirus and enterotoxigenic *E. coli* are common. Others include *Shigella*, rotavirus, *Salmonella*, *Giardia* (a parasite), and *Campylobacter*. In real life, if you get diarrhea while traveling, it's traveler's diarrhea, regardless of the cause.

Q: What is the cause of amebic meningoencephalitis?

A: *Naegleria fowleri* is the primary cause. The disease is rare and usually fatal.

Q: Describe the hemolytic pattern of *Clostridium perfringens* on a blood agar plate.

A: It has a double zone of hemolysis. The inner zone is clear (beta-hemolysis) and the outer zone is partial. The partial hemolysis is a light red or pink (not the typical green as in alpha-hemolysis).

Q: How do bacteria release exotoxins and endotoxins (what is the mechanism)?

A: Exotoxins are typically secreted. Gram-negative bacterial cell walls are more complex than those of gram-positives and require sophisticated bacterial secretion systems. One classification scheme includes six of these systems, designating each as types I through VI. Types III and IV are particularly important in bacterial virulence. Endotoxins are released after cell lysis of the gram-negative bacteria. They are not present in gram-positive cell walls.

Q: List the Reiter syndrome triad and associated infections.

A: Reiter syndrome is the triad of urethritis, conjunctivitis and uveitis, and arthritis. It is an autoimmune disease that has a strong association with HLA type B27. *Chlamydia trachomatis* is the most common infectious agent that can lead to Reiter syndrome. GI infection caused by *Salmonella*, *Shigella*, *Campylobacter*, or *Yersinia enterocolitica* may also lead to Reiter syndrome.

Q: The therapy for some infections, such as tetanus, requires the use of antitoxin. What is antitoxin and what does it do?

A: Antitoxin is another term for a specific immunoglobulin. It is a preformed antibody that when injected into the body binds to and inactivates the specific toxin to which it was made. It binds to and inactivates only circulating toxin. Any toxin that has already bound to its cellular receptor is unaffected.

Q: What are beta-lactamase inhibitors?

A: When used in combination with a beta-lactam antibiotic such as amoxicillin, beta-lactamase inhibitors "convert" a beta-lactam–resistant organism into a susceptible organism. The resistant organism must be resistant because of the secretion of beta-lactamase. Beta-lactamase inhibitors bind irreversibly to the beta-lactamase. This inactivates the beta-lactamase enzyme so that it cannot hydrolyze the beta-lactam antibiotic. The beta-lactamase inhibitors currently in use are clavulanic acid (used in combination with amoxicillin or ticarcillin), tazobactam (used in combination with piperacillin), and sulbactam (used in combination with ampicillin).

Q: A 38-year-old male drug addict was found unconscious by his girlfriend. He was taken to the emergency room and admitted to the hospital. He later developed fever, cough, and foul-smelling breath. What is the most likely diagnosis?

A: Aspiration pneumonia, most likely due to anaerobes. Patients who lose consciousness can aspirate from the oral cavity and stomach contents. The major normal flora bacteria in the oral cavity are the anaerobes. Always consider anaerobes if a foul-smelling odor is associated with an infection.

Q: What kinds of cells can be infected with HIV and what is the receptor?

A: Helper T cells (T_H) are the primary targets, but other cells such as macrophages that are CD4+ can also be infected.

Q: What are the three major forms of Hansen disease?

A: Leprosy is called Hansen disease in the United States. Tuberculoid, lepromatous, and borderline are the major forms. Intermediary forms (borderline tuberculoid and lepromatous) also occur. About 100 cases are reported annually in the United States.

Q: A 21-year-old college student presented to student health because of a severe sore throat and a temperature of 102°F (38.9°C). On physical examination, the doctor also observed lymphadenopathy. There was a whitish coating on the tonsils. The student was heterophil antibody positive. A preliminary diagnosis of a viral infection was made. Where in infected cells does this virus replicate?

A: Nucleus. This is infectious mononucleosis due to EBV, which is a herpesvirus. Herpesviruses are DNA viruses that bud through the nuclear membrane. Of the DNA viruses, only poxviruses do not replicate in the nucleus. Of the DNA viruses, only herpesviruses bud through the nucleus.

Q: Where do most RNA viruses replicate?

A: Most replicate in the cytoplasm

Q: Which RNA viruses do not replicate in the cytoplasm?

A: Retroviruses and influenza viruses

Q: List some human virus–cancer associations.

A: EBV–Burkitt lymphoma and nasopharyngeal carcinoma; hepatitis B virus–hepatocellular carcinoma; papilloma viruses–cervical carcinoma; HTLV-I–adult T-cell leukemia; HHV-8–Kaposi sarcoma

Q: List some human bacteria–cancer associations.

A: *Helicobacter pylori*–adenocarcinoma of the stomach. This is the only known one so far.

Q: What gene of HIV-1 encodes for reverse transcriptase?

A: The polymerase (*pol*) gene, which also encodes for the protease and integrase enzymes

Q: What is the function of the reverse transcriptase?

A: Reverse transcriptase uses the HIV RNA as a template to synthesize DNA in infected cells.

Q: What are the functions of the HIV protease and integrase?

A: The HIV protease cleaves viral precursor proteins into smaller viral proteins that are part of the structure of the HIV virions. This enzyme is inhibited by protease inhibitors. The integrase allows the DNA (transcribed from the viral RNA) to integrate into the host cell DNA.

Q: What pathogen carries the highest risk of transmission by blood transfusion?
A: Hepatitis C

Q: There is simultaneous activation of coagulation and fibrinolysis in what pathological condition?
A: Disseminated intravascular coagulation

Q: What type of bacteria is most commonly cultured from brain abscesses: aerobic or anaerobic?
A: Anaerobic, by contiguous spread from the sinuses and oral cavity

Q: What bacteria most commonly cause meningitis in infants younger than 1 year of age?
A: *Streptococcus agalactiae* and *E. coli*

Q: What is the treatment for *Listeria monocytogenes* in a newborn?

A: Ampicillin (drug of choice) alone or with gentamicin. Other drugs include trimethoprim-sulfamethoxazole, erythromycin, or penicillin G. These are also used for older children and adults.

Q: A heat-stable (100°C) bacterial exotoxin that is elaborated in contaminated foods in sufficient quantity to cause disease without the multiplication of the causative bacterium in the human gut is associated with what organism?

A: The toxin described is the enterotoxin elaborated by *Staphylococcus aureus*. The heat-stable enterotoxin of *Escherichia coli* must be synthesized and secreted in the host's intestine.

Q: The usual source of the causative agent of pseudomembranous colitis is what?

A: *Clostridium difficile* is a normal inhabitant of the colon in many people; therefore most infections are from one's own normal flora. Some cases may be nosocomially spread by spores from a patient with active disease. Although typically a nosocomial infection, community-acquired disease has been noted with increasing frequency.

Q: What are the first and second most common complications of AIDS?
A: First, *Pneumocystis jiroveci* (*carinii*) pneumonia; second, Kaposi sarcoma

Q: An AIDS patient presented with the signs and symptoms of CNS cryptococcal infection (headache, depression, lightheadedness, seizures, and cranial nerve palsies). How is the diagnosis confirmed?

A: A diagnosis is confirmed by India ink prep, a fungal culture, or by testing for the presence of cryptococcal antigens in the CSF.

Q: An AIDS patient presented with symptoms suggesting tuberculous meningitis (fever, meningismus, headache, seizures, focal neurological deficits, and altered mental status). What is the most likely organism?

A: A member of the *Mycobacterium avium* complex (MAC). There are several subspecies in the complex. These bacteria were previously referred to as *Mycobacterium avium-intracellulare*.

Q: What is the most common cause of retinitis in AIDS patients?

A: CMV. Findings include photophobia, redness, scotoma, pain, or a change in visual acuity. Fluffy white retinal lesions may be evident.

Q: EBV has been associated with what malignancies?

A: Nasopharyngeal carcinoma, Burkitt lymphoma, and Hodgkin lymphoma

Q: Hepatitis B has been associated with what cancer?

A: Hepatocellular carcinoma

Q: What is the most common GI complaint in AIDS patients?

A: Diarrhea. Hepatomegaly and hepatitis are also typical. Conversely, jaundice is an uncommon finding. *Cryptosporidium* and *Isospora* are common causes of prolonged watery diarrhea.

Q: What are the target cells and vehicles for dissemination of HIV in the infected host?

A: CD4+ T cells (target cells) and monocytes, macrophages, and dendritic cells (dissemination)

Q: When should prophylaxis against *Pneumocystis jiroveci* (*carinii*) pneumonia be initiated in HIV patients?

A: When CD4+ lymphocyte counts fall below 200. Therapy can be discontinued for patients receiving highly active antiretroviral therapy with an increase in CD4+ count to >200 cells/mm^3 for at least three months.

Q: What is the cause of oral hairy leukoplakia?

A: EBV

Q: A 47-year-old man was admitted for an upper GI bleed from a gastric ulcer. The laboratory reported a positive urease test based on a breath test. What is the suggested therapy?

A: This is a case of *Helicobacter pylori* infection. It is important to know that monotherapy is not indicated. Current recommendations include a combination of a proton pump inhibitor + clarithromycin + either metronidazole or amoxicillin. An alternative is bismuth subsalicylate + metronidazole + tetracycline + either a proton pump inhibitor or H$_2$ blocker.

Q: What parasites involve the human lung as part of the infectious process?

A: Larvae of *Ascaris lumbricoides* migrate through the lungs (ingestion of larvae → small intestine → portal venules → liver → hepatic vein → heart → lung → coughed up and swallowed → small intestine) resulting in respiratory symptoms. The lung fluke (*Paragonimus*) reaches maturity in the lungs after the metacercariae encyst in the duodenum and later migrates from that site. Hookworm larvae (*Necator americanus* and *Ancylostoma duodenale*) may migrate through the lungs after skin penetration.

Q: A patient who developed scarlet fever had a previous infection with what organism?

A: *Streptococcus pyogenes*

Q: What organism is the most common cause of UTI?

A: *Escherichia coli*

Q: Which species of malaria parasite has the highest mortality in human infections?

A: *Plasmodium falciparum*

Q: A patient in a rural area of Puerto Rico developed symptoms (e.g., fever, chills, cough, diarrhea, splenomegaly, abdominal pain, rash, plus others) that suggested a parasitic infection. Examination of the stool revealed large oval eggs that had a prominent lateral spine. What is the most likely organism?

A: The blood fluke *Schistosoma mansoni*. Of the several species of *Schistosoma*, *S. mansoni* is the only one found in the Western Hemisphere.

Q: A 6-year-old boy developed a rash, headache, and high fever (104°F) several days after returning from a hike in a wooded area of North Carolina. The rash began on his hands and feet, including the palms and soles, and spread toward the trunk. How did the child most likely acquire this disease?

A: Tick bite. This is a classic case of Rocky Mountain spotted fever.

Q: A 7-year-old girl developed enlarged lymph nodes after being scratched by a neighborhood cat. What is the most probable organism?

A: *Bartonella henselae*

Q: DNase is an enzyme that can degrade free DNA that is in solution. What type of genetic transfer between bacteria can be prevented if DNase is put into the solution?

A: Transformation. In transformation, the DNA is free in solution before taken up by bacteria.

Q: Which age group usually contracts croup (laryngotracheobronchitis)?

A: Young children. Croup is characterized by cold symptoms, a sudden barking cough, inspiratory and expiratory stridor, and a slight fever.

Q: What causes croup?

A: Primarily parainfluenza virus

Q: What is the treatment for individuals infected with *Salmonella*?

A: Supportive care without antibiotics. However, if a severe fever is exhibited, antibiotic therapy may be warranted. The use of antibiotics can occasionally induce a carrier state and can select for antibiotic resistant strains of normal flora bacteria. Salmonellosis is usually self-limited.

Q: Empyema is most often caused by which organism?

A: *Staphylococcus aureus*. Gram-negative organisms and anaerobic bacteria also cause empyema.

Q: Which age group typically gets epiglottitis?

A: Young children, although any group can get epiglottitis

Q: What are the most common causes of otitis media?

A: The most common pathogens are *Streptococcus pneumoniae*, non-typeable *Haemophilus influenzae*, *Moraxella catarrhalis*, and viruses such as rhinovirus and adenovirus.

Q: What is the difference between the "cough of croup" and the "cough of epiglottitis"?

A: Croup has a seal-like "barking" cough, whereas epiglottitis is accompanied by a minimal cough. Children with croup have a hoarse voice, whereas those with epiglottitis have a muffled voice.

Q: What antibiotics should be used to treat acute purulent rhinitis?

A: Although antibiotics may be useful for treatment, the general recommendation for both purulent and clear rhinitis is to not treat with antibiotics. Symptoms of rhinitis may last up to two weeks.

Q: Should a 7-year-old who has never received her Hib vaccine be vaccinated now?

A: No, most children are immune by the age of 5.

Q: Sin Hombre virus occurs most commonly in what geographical location?

A: In the southwestern United States, but it can be found in other states as well. It is a hantavirus carried by rodents, especially deer mice.

Q: Where is the endemic area for histoplasmosis?

A: The Ohio and the Mississippi River valleys. Nearly 100% of the population is seropositive in endemic areas. However, only 1% of these individuals develop the active disease.

Q: What disease should be suspected in a patient with a two-week history of lower limb weakness?

A: Guillain-Barré syndrome usually causes an ascending weakness that begins in the lower extremities. Conversely, the weakness is descending with botulism poisoning. Cranial nerves are typically affected first with myasthenia gravis. About half the cases of Guillain-Barré syndrome in the United States are due to sequelae after infection with *Campylobacter jejuni*. There can be other infectious and noninfectious etiologies.

Q: What types of bacteria are most commonly found in lung abscesses?

A: Anaerobic bacteria, which are normal flora of the oral cavity. Aspiration of these bacteria into the lungs can lead to aspiration pneumonia.

Q: What is the most common cause of pneumonia acquired in the hospital?

A: Gram-negative bacilli are most common, especially *Pseudomonas aeruginosa*. Other common causes include *Klebsiella*, *Enterobacter*, *Serratia*, and *Acinetobacter*. There is a low probability of *Serratia* or *Acinetobacter* appearing as a correct answer on a standardized exam. Nosocomial pneumonias due to *Staphylococcus aureus* are also common.

Q: What is the most common community-acquired pneumonia due to a gram-negative rod?

A: You might see different answers to this question, but the most common is probably *Pseudomonas aeruginosa*. Some sources say *Klebsiella pneumoniae*.

Q: What is the most common cause of upper respiratory tract infections: bacteria or viruses?

A: Viruses account for most cases of pharyngitis, laryngitis, and bronchitis.

Q: What secondary bacterial pneumonia often occurs after a viral pneumonia?

A: *Staphylococcus aureus* pneumonia

Q: What are the classic signs and symptoms of TB?

A: Night sweats, fever, weight loss, malaise, cough, and greenish-yellow sputum (most commonly seen in the mornings)

Q: Is an upper respiratory tract infection accompanied by a high fever usually caused by bacteria or viruses?

A: Bacteria. The high fever is the clue. However, most upper respiratory tract infections are due to viruses.

Q: How does bacterial meningitis differ from viral meningitis in terms of the corresponding CSF lab values?

A: Bacterial meningitis will have low glucose and high protein levels, whereas viral meningitis will have normal glucose and a normal protein.

Q: A 31-year-old man develops fever, sinus pain, and bloody nasal drainage several weeks after undergoing bone marrow transplantation for leukemia. Branching hyphae are seen in a silver stain of the biopsy. What is the most likely pathogen?

A: *Aspergillus*

Q: List some diseases of CMV in immunocompromised patients.

A: Retinitis, pneumonitis, and hepatitis

Q: Why is attenuated *Mycobacterium bovis* (bacille Calmette-Guérin, or BCG) not routinely used as a TB vaccine in the United States?

A: The TB rate in the United States is low in the general population. The tuberculin skin test is used to identify possible infections. People vaccinated with BCG have a positive skin test, making it difficult to identify new cases. However, interferon-gamma release assays are not affected by previous BCG vaccination and may help to eliminate the unnecessary treatment of persons with false-positive results.

Q: Koplik spots are pathognomonic for what infection?

A: Measles

Q: List the major enveloped viruses.

A: **DNA viruses:** Herpesviruses (herpes simplex, CMV, EBV, varicella-zoster virus), hepatitis B, variola (smallpox)

RNA viruses: RSV, measles, mumps, rubella, influenza, parainfluenza, HIV, HTLV, and rabies. Note that most enveloped viruses contain RNA.

Q: List the major nonenveloped (naked) viruses.

A: **DNA viruses:** Papillomaviruses, polyomaviruses, adenoviruses, parvoviruses (B19 [fifth disease])

RNA viruses: Enteroviruses (hepatitis A, coxsackievirus, echovirus, poliovirus), rotavirus, rhinovirus

Q: What virus is the major cause of lower respiratory tract infections in infants?

A: RSV

Q: A 54-year old man presents to your office complaining of a painful right index finger. The finger is swollen on one side of the nail and is erythematous, warm, and quite tender. Microscopy of an aspirate reveals gram-positive cocci in clusters. What should be considered for empiric antimicrobial therapy?

A: A penicillinase-resistant beta-lactam. The etiologic agent is probably *Staphylococcus aureus*. Therapy would need to be modified if susceptibility testing demonstrates that the organism is methicillin resistant. Drainage should also be considered.

Q: Metronidazole is a drug of choice for what kinds of infections?
  A: Anaerobic, especially due to *Bacteroides fragilis* (and related bacteria); *Clostridium difficile* and *C. tetani*; protozoal infections, especially trichomoniasis, giardiasis, and amebiasis; and *Helicobacter pylori* (in combination with clarithromycin and a proton pump inhibitor or in combination with tetracycline, bismuth subsalicylate, and either a proton pump inhibitor or H_2 blocker)

Q: What fungus is commonly associated with subcutaneous infection after a puncture wound due to a rose thorn?
  A: *Sporothrix schenckii*, although you can get this infection from other sources

Q: Who is at most risk for *Aspergillus fumigatus* infection?
  A: Immunocompromised patients, in whom it frequently presents as a lung infection. Filamentous septated branching hyphae can be seen microscopically in specimens.

Q: What dimorphic fungus found primarily in the Mississippi and Ohio River valleys is associated with the droppings of birds and bats and can cause serious life-threatening infections?
  A: *Histoplasma capsulatum*. The bird droppings (particularly blackbirds and chickens) and bat guano (especially in caves) in the soil support luxuriant growth of the *Histoplasma*.

Q: What organism most commonly causes acute bacterial endocarditis?
  A: *Staphylococcus aureus*

Q: What is the approximate mortality rate of pulmonary anthrax after the patient becomes symptomatic?
  A: Nearly 100%. That is one reason it is considered a good bioterrorist weapon.

Q: List some gram-negative bacteria that can cause endocarditis.
  A: The HACEK group (*Haemophilus, Actinobacillus, Cardiobacterium, Eikenella,* and *Kingella*) can be causes of culture-negative endocarditis. Many other gram-negatives have been associated with endocarditis, including *Neisseria, Pseudomonas, Serratia* (and other enterics), and *Bartonella*. Yeasts, especially *Candida*, can also be a cause.

Q: How is CMV classified?
  A: It is a herpesvirus.

Q: What is the most common agar medium that is used for the isolation of fungi from clinical specimens?
  A: Sabouraud dextrose medium

Q: In a microbiology genetics experiment, it was found that a gene coding for ampicillin resistance could be transferred from one strain of *E. coli* to another strain when both strains were mixed together in nutrient broth. On further investigation, it was found that if the two strains were kept separated in the broth by a filter (0.45 μm pore size), gene transfer did not occur. What is the mechanism of this genetic exchange?
  A: Conjugation. The filter keeps the cells from making contact. Transformation is wrong because in transformation the pieces of DNA would be able to pass through the filter. Similarly, transduction is wrong because in transduction the bacteriophages would be able to pass through the filter.

Q: Some strains of *Streptococcus pyogenes* can produce a pyrogenic toxin that may result in scarlet fever. Research has shown that a toxin gene carried by a bacteriophage codes for this toxin. What is the name of this genetic phenomenon?
  A: Lysogenic conversion

Q: In the exchange of DNA between two different strains of *E. coli* by transformation, is the cell that donates the DNA designated as an F⁺?
  A: No. Transformation is the uptake of free DNA that was released by any donor cell when it lysed. The recipient cell can potentially take up soluble chromosomal or plasmid DNA from any cell.

Q: A 67-year-old retired plumber was in a serious automobile accident and was admitted to the hospital after being treated in the emergency room. He had head trauma as well as a compound fracture of the femur. Four days after admission, the open wound on the leg was showing signs of infection. Especially noticeable were the necrosis, gas production, and foul smell. On subsequent culture, a double zone of hemolysis was observed on a blood agar plate. What would the bacteria look like microscopically?
  A: Gram-positive "boxcar"-shaped rods. This is a description of an anaerobic infection (necrosis, gas production, and foul smell) due to *Clostridium perfringens* (double zone of hemolysis).

Q: From the previous question, what is the most appropriate antimicrobial agent?

A: Penicillin. Metronidazole should also be considered and might be your only choice on a standardized exam.

Q: *Klebsiella pneumoniae* is a gram-negative, nonmotile rod that ferments lactose. It possesses a large antiphagocytic polysaccharide capsule and is urease positive. Can this organism cause urinary calculi?

A: Yes, because it is urease positive

Q: A 16-year-old boy had been hospitalized due to a gun shot wound to the lower abdomen that perforated the ascending colon. One day after admission he became febrile (39.8°C [103.7°F]). Blood culture resulted in the isolation of a gram-negative anaerobic rod that was resistant to penicillin but susceptible to metronidazole and clindamycin. What is the most probable identification of the organism?

A: *Bacteroides fragilis*. It constitutively secretes a beta-lactamase.

Q: What is the relationship between cystic fibrosis and biofilms?

A: *Pseudomonas aeruginosa* and other bacteria involved in cystic fibrosis grow in the respiratory tract as thick tenacious biofilms.

Q: A 21-year-old woman developed diarrhea the day after eating at a local family-owned restaurant where she had eaten tacos made with ground beef. The diarrhea lasted for four days and resolved without the use of antimicrobial therapy. Stool cultures resulted in the isolation of a motile, gram-negative, non–lactose-fermenting bacillus that was hydrogen sulfide positive. What organism is the most probable pathogen?

A: *Salmonella enterica*

Q: Gram-negative bacilli that are resistant to multiple antimicrobials frequently contain resistant transfer factors, or RTFs. What is an RTF?

A: An RTF is a plasmid that has the genes necessary to code for the transfer of itself by conjugation to another cell and has at least one gene that codes for antibiotic resistance. If there is more than one gene that codes for resistance, then it is multiply resistant. These cells can be F+ or Hfr.

Q: What is an Hfr cell?

A: Hfr stands for high frequency of recombination. It is a male cell in which the F plasmid has integrated into the bacterial chromosome. During conjugation, the bacterial chromosome is transferred before the integrated plasmid is transferred. Because of the high homology, the DNA from the donor can recombine into the recipient chromosome at a high frequency.

Q: What is an F' (F prime) cell?

A: It is a male cell in which the F plasmid first integrates into the bacterial chromosome. Subsequently, when the integrated F plasmid comes back out of the chromosome, it carries a piece of the chromosome and the cell is now called F'. An F' cell can function as a male in conjugation.

Q: In analyzing the genotypes after transduction experiments, it was observed that all the transduced cells had genetic sequences from the donor cells that came from the same general area (within a locus spanning no more than two genes) of the chromosome. Why?

A: This is an example of specialized transduction. One of the best examples is transduction using lysogenic cells of *E. coli* that are infected with bacteriophage lambda. The lambda phage inserts into the bacterial chromosome only at a specific site (between the *bio* and *gal* genes). On induction and subsequent lysis of the bacterial cells, most virus particles contain viral DNA, but some may contain bacterial chromosomal DNA from either the adjacent *bio* or *gal* genes.

Q: An anaerobic gram-negative rod was isolated from an oral lesion of a 24-year-old female patient. What is the drug of choice?

A: For almost all anaerobic infections, penicillin is your best guess. Metronidazole or clindamycin also works well for most anaerobic infections (especially *Bacteroides fragilis*, *Clostridium difficile*, and *C. tetani*). Never choose penicillin (unless you also include a beta-lactamase inhibitor) for infections due to *Bacteroides fragilis* or related bacteria (they produce a beta-lactamase). In the present case, *B. fragilis* would not be suspected because it is normal flora of the lower intestinal tract, not the oral cavity. Therefore penicillin should work. In real life, metronidazole or clindamycin are often the drugs of choice in suspected anaerobic infections.

Q: The physician of a patient with endocarditis requested the value for the MBC of the antibiotic she was using. What method of susceptibility testing can be used to determine this value?

A: Broth dilution method. After the MIC is determined by this method, broth (i.e., growth medium) from the tubes with no growth is transferred to fresh broth containing no antibiotic. These tubes are incubated to see if anything grows. The first tube with no growth is the MBC.

Q: List the Centers for Disease Control and Prevention Category A bioterrorism agents and the diseases they cause.

A:
- *Bacillus anthracis*: anthrax
- *Clostridium botulinum* toxin: botulism
- *Francisella tularensis*: tularemia
- *Yersinia pestis*: plague
- Hemorrhagic fever viruses: filoviruses (e.g., Ebola, Marburg) and arenaviruses (e.g., Lassa, Machupo)
- Variola major virus: smallpox

Q: *Mycobacterium tuberculosis* is unusually sensitive to the killing effect of UV light. The primary effect of UV light on this organism is what?

A: Thymine dimer formation

Q: What kind of chemical bonds form between the two thymines of the thymine dimer?

A: Covalent

Q: Penicillin can eventually lead to the death of susceptible bacteria. Before this happens, the penicillin must first bind to what enzyme?

A: Transpeptidase

Q: What is the most common organism in human feces?

A: *Bacteroides* is the best choice. Although *Bacteroides fragilis* is not the most common (another species in the *Bacteroides fragilis* group [*B. vulgatus*] is the most common), it will probably be the best answer if it is a choice. Most fecal organisms are obligate anaerobes (>99.9%).

Q: What is the most common manifestation of neurosyphilis?

A: Tabes dorsalis. It is usually seen in tertiary syphilis and is due to a slow and progressive degeneration of nerve cells and fibers in the spinal cord. It is characterized by an unsteady gait, painful paresthesias (aka lightning pains), and urinary incontinence.

Q: If used as a bioterrorist weapon, what would be the optimal method of delivery for *Bacillus anthracis*?

A: Aerosolized spores

Q: How is *Pneumocystis jiroveci* (*carinii*) classified?

A: Although it looks like a parasite, it is actually a fungus.

Q: A research scientist had developed what he hoped was a new biopolymer that could be used as a catheter or an implant and that would resist colonization by bacteria. The polymer could be easily molded into different shapes at 65°C but became brittle and cracked above 75°C. How can this polymer be sterilized?

A: Ethylene oxide or beta-propiolactone gases are good choices. Other methods include gamma rays or x-rays.

Q: List some of the biological activities of the LPS of gram-negative bacteria.

A: Activation of complement, hypotension, disseminated intravascular coagulation, and fever are some examples. Almost all the biological activities are due to the endotoxin-mediated (i.e., lipid A) release of various cytokines, including IL-1, IL-6, IL-8, TNF, and platelet-activating factor. Superantigens have a similar effect.

Q: What specifically is the effect of penicillin on peptidoglycan (i.e., what is the specific target and what happens to the cell)?

A: Penicillin and other beta-lactam antibiotics (cephalosporins, cephamycins, carbapenems, monobactams, etc.) inhibit the transpeptidase enzyme. Transpeptidase (penicillin-binding protein) is the last enzyme involved in peptidoglycan synthesis. The enzyme connects, via covalent bonds, the no. 4 amino acid of one tetrapeptide side chain to the no. 3 amino acid of another tetrapeptide side chain. This connection can be a direct covalent bond between the amino acids, or it can be a peptide cross-bridge between the two amino acids. Either way, the result after growth in the presence of a beta-lactam antibiotic is an inhibition of peptidoglycan synthesis (an incompletely synthesized peptidoglycan, i.e., no connections between tetrapeptide cross bridges). Autolytic enzymes within the cells then break down the remaining peptidoglycan, resulting in cell lysis and death. Notice that a major difference between lysozyme and beta-lactam antibiotics is that lysozyme digests peptidoglycan and beta-lactams stop its synthesis.

Q: What is an antimetabolite that competitively inhibits the formation of folic acid?

A: The sulfonamide group most readily comes to mind, but others include dapsone, sulfoxone, and aminosalicylate.

Q: What causes bad breath?

A: Anaerobes in the oral cavity

Trivia: Recent evidence suggests the gram-positive anaerobe *Solobacterium moorei* is the primary culprit. This is a rare, but documented, human pathogen.

Q: How do humans get infected with *Borrelia recurrentis*?

A: Bites from body lice; thus the disease is also called louse-borne relapsing fever. Louse-borne relapsing fever is found in Africa and South America. Other species of *Borrelia* cause tick-borne relapsing fever, which is found in the United States.

Q: What causes relapsing fever in the United States?

A: Primarily *Borrelia hermsii*. The disease is carried by soft-bodied ticks and is most common in western states. Commonly, there is only a tick bite, but no tick. The tick comes out at night, usually when campers are sleeping in a cabin, takes a blood meal, then hides until the next meal becomes available. The ticks are attracted to carbon dioxide exhaled by the sleeping campers.

Q: A 67-year-old alcoholic was found in an alley, covered in his own vomit and beer. Upon examination, he is shaking, has a fever of 103.5°F, and is coughing up currant jelly sputum. What is the diagnosis?

A: Pneumonia due to *Klebsiella pneumoniae*. The key here is the sputum resembling currant jelly. Always associate this with *K. pneumoniae*, especially in alcoholics. One study showed a 100% mortality rate. Note: These severe signs and symptoms take longer to develop than the scenario implies.

Q: Along these same lines, what is the most common cause of community-acquired pneumonia in alcoholics?

A: Typically, the answer they are looking for is *K. pneumoniae*; however *Streptococcus pneumoniae* is the number one cause. To get the question right, look for additional information in the stem of the question: gram-positive diplococci (*S. pneumoniae*), current jelly (*K. pneumoniae*), aspiration pneumonia (anaerobes, usually mixed with gram-negative Enterobacteriaceae), male older than 55 and attending a convention (*Legionella*; there may be other clues here such as no growth on typical laboratory media or no bacteria seen on Gram stain).

Q: What are some microbial causes of myocarditis?

A: Viruses in general, especially coxsackieviruses, but influenza and other viruses can cause this. Other etiologies include postviral myocarditis, an autoimmune response to recent viral infection, bacteria (diphtheria and tuberculosis), fungi, protozoa (Chagas disease), and spirochetes (Lyme disease).

Q: *Streptococcus bovis* isolated from the blood of a patient may be indicative of what noninfectious disease?

A: The patient may have a malignant or premalignant GI tract lesion, usually colonic.

Q: What is the hemolytic pattern of the viridans streptococci?

A: They are alpha-hemolytic (green).

Q: Describe the rash of scarlet fever.

A: The bright red non-itchy rash on the skin begins as small red macules, especially on the cheeks and neck and in the groin area. The temples and cheeks may be red, but the area around the nose and mouth remains pale. The rash normally spreads from the chest to the trunk and extremities. It may feel rough (sandpaper-like) as it fades. Although the patient might have "strawberry tongue," this sign can also be seen in Kawasaki disease.

Q: Within eight hours after eating reheated rice, a 25-year-old medical student developed abdominal cramps and vomiting. What is the most likely organism responsible?

A: *Bacillus cereus*. Two clues: reheated rice and rapid onset. *B. cereus* food poisoning can present as either vomiting or diarrhea (or both) depending on the type of toxin produced. *B. cereus* food poisoning is not unique to rice. It can be associated with other foods, such as pasta, meat, and vegetable dishes, especially when held at warm (not hot) temperatures for several hours.

Q: What two other bacterial food poisonings have a relatively rapid onset?

A: *Staphylococcus aureus* (preformed toxin in food) and *Clostridium perfringens* (toxin produced in large intestine as bacteria sporulate)

Q: What is the usual source of bacteria in aspiration pneumonia?

A: The oropharynx and stomach (gastric contents). Aspiration of oral microbial flora can lead to infectious pneumonia by a more direct route than aspiration of stomach contents. Aspiration of stomach contents can cause chemical (acid) damage so that aspirated bacteria can grow. This is not required for oral flora.

Q: What are the most common causes of acute viral pneumonias?

A: Influenza virus, adenovirus, parainfluenza viruses, and RSV. Most cases of pneumonia in children and infants are due to viruses. RSV and the bacterium *Chlamydia trachomatis* are most commonly seen in infants.

Q: Which virus listed in the previous question is most likely to cause conjunctivitis?

A: Adenoviruses, from contaminated water such as swimming pools, or from fomites. There are two primary types of viral conjunctivitis. Pharyngoconjunctival fever is usually due to adenovirus 3 and is characterized by fever, sore throat, and follicular conjunctivitis. Adenovirus types 8 and 19 usually cause epidemic keratoconjunctivitis. Both types of conjunctivitis are highly contagious.

Q: List the uses of KOH in the diagnosis of infectious diseases.

A: KOH preparations of skin scrapings, tissue aspirates, sputum, histologic sections, and other specimens reveal the hyphae of dermatophytes and other fungi, spherules of *Coccidioides*, and yeast cells of *Candida*, *Blastomyces*, and so on. Addition of KOH to vaginal secretions (whiff test) in suspected cases of bacterial vaginosis can be useful in diagnosis (causes release of aromatic amines resulting in a fishy odor).

Q: What characteristics can be used for the identification of *Candida albicans*?

A: *C. albicans* is a typical budding yeast that can produce pseudohyphae and germ tubes. Other tests can be done, but you do not need to know them.

Q: Some viruses can be grown in the laboratory in cell cultures. What is the general name given to the morphologic change in the cells that is indicative that the cells are infected?

A: Cytopathic effect (sloughing of cells from surface, rounding of cells, giant cells, etc.)

Q: An acid-fast stain was performed on a stool specimen from an AIDS patient who had diarrhea. Large (5–7 μm) acid-fast spherical cells were observed. The most likely organism is what?

A: *Cryptosporidium parvum*. This is the most common cause, but at least six other *Cryptosporidium* species can cause diarrhea even in immunocompetent individuals.

Q: A Gram stain from a positive blood culture bottle showed gram-positive cocci. Subculture to a blood agar plate grew an organism that was catalase positive and coagulase negative. What was the organism?

A: Any *Staphylococcus* except *S. aureus*. If it is a gram-positive coccus, it is most likely *Staphylococcus* or *Streptococcus*. It does not matter that the arrangement (clusters or chains) of the cells is not given because the catalase test will differentiate between the two (*Staphylococcus* is catalase positive and *Streptococcus* is catalase negative). Of the staphylococci, only *S. aureus* is coagulase positive.

Q: What common pathogen produces red colonies on MacConkey agar?

A: Any lactose-fermenting gram-negative bacilli. There can be several species (*Klebsiella*, *Enterobacter*, *Citrobacter*, *Serratia*, *Hafnia*, *Pantoea*, *Yersinia*, others), but the most common is *E. coli*.

Q: In a microbiology research lab, it was found that a strain of bacteria produced a particular enzyme only if it was infected with a specific bacteriophage. This is an example of what phenomenon?

A: Lysogenic conversion. The gene coding for the new phenotype (in this case, a new enzyme) is located on the virus DNA.

Q: In the previous question, what general name is given to the virus DNA?

A: Prophage. A prophage is the bacteriophage DNA that inserts itself into the host chromosome. In human and animals a similar phenomenon occurs and the integrated virus nucleic acid is called a provirus. Retroviruses (i.e., their transcribed DNA) are examples.

Q: From the above two questions, what name is given to bacteriophages that can stably insert themselves (their DNA) into the host chromosome after infection?

A: Lysogenic or temperate bacteriophage. Bacteriophage lambda is an example.

Q: What is the gene of HIV-1 that encodes reverse transcriptase?

A: The *pol* (polymerase) gene

Q: What are the target cells of HIV in the infected host?

A: CD4+ T cells, monocytes, macrophages, alveolar macrophages, microglial cells (brain), and dendritic cells

Q: What is the species name of the tapeworm that causes cysticercosis?

A: *Taenia solium*

Q: From the previous question, what is the common name?

A: Pork tapeworm. In the United States most cases are imported. Infection is usually asymptomatic, but in rare cases it can have serious manifestations, including neurocysticercosis.

Q: What organism is the most common cause of nosocomial UTIs?

A: *E. coli*. It is also the most common cause of community-acquired UTIs.

Q: What are some organisms that can cause gastroenteritis with bloody diarrhea?

A: Bacteria such as *Campylobacter jejuni, Shigella dysenteriae, E. coli* O157:H7, and *Salmonella enterica* can cause bloody diarrhea. Viral gastroenteritides are not bloody.

Q: What are some infections due to *Aeromonas*?

A: *Aeromonas*, especially *A. hydrophila*, is an emerging pathogen causing several diseases, including gastroenteritis; hemolytic uremic syndrome; urine, wound, and burn infections; necrotizing fasciitis; pneumonia; meningitis; and eye infections. It is a gram-negative, oxidase-positive organism, thus more closely related to *Pseudomonas* than *E. coli*. They are common in many environments but are best known for being found in water, even if chlorinated.

Q: *Nocardia asteroides* is a gram-positive rod that is also partially acid-fast. Where do primary infections occur in the body?

A: Lung. *N. asteroides* is found in soil. Humans are infected by inhalation.

Q: Where do secondary infections of *N. asteroides* occur?

A: *N. asteroides* can disseminate from the lungs to other organs, including the brain.

Q: An AIDS patient developed a severe watery diarrhea. A modified acid-fast stain demonstrated spherical acid-fast organisms in the stools. What is the organism and how is it classified?

A: *Cryptosporidium* is a parasite (a protozoan). The oocytes seen in stools are acid-fast. Useful therapy includes nitazoxanide (a thiazolide compound active against some parasites, including *Giardia duodenalis* [syn. *G. lamblia* or *G. intestinalis*]), paromomycin (taken orally; a nonabsorbable aminoglycoside), and azithromycin. These drugs reduce the duration of diarrhea but commonly do not lead to a cure. Fluid and electrolyte therapy is critical.

Q: What characteristic of bacteria can lead to urinary calculi in a UTI?

A: Although there can be other mechanisms, urease production is often important.

Q: Name two urease-positive bacteria that are common causes of UTIs.

A: *Proteus* and *Klebsiella*. You do not need to memorize the different species. Assume they are all the same when it comes to signs and symptoms, identification, and so forth.

Q: What bacterium causes food poisoning because of enterotoxin production during sporulation in the gut?

A: *Clostridium perfringens*

Q: What bacteria produce a heat-stable enterotoxin in food?

A: The two that you should know are *Staphylococcus aureus* and *E. coli*. In traveler's/infant diarrhea due to enterotoxigenic *E. coli*, one of the enterotoxins is heat stable.

Q: Although pregnant females generally show mild symptoms after infection, what food-borne bacterium can cause serious life-threatening infection of the fetus or neonate?

A: *Listeria monocytogenes*

Q: What infectious agents can pets transmit?

A: *Toxoplasma gondii* (cats; toxoplasmosis), *Ancylostoma braziliense* (cats and dogs; hookworm; cutaneous larva migrans), *Toxocara canis* (dogs; roundworm; visceral and ocular larva migrans), rabies virus (most commonly in emerging countries with no or limited pet vaccinations), *Pasteurella multocida* (infection after cat or dog bite), *Chlamydophila psittaci* (wild and domestic birds, including parakeets, etc.), and *Campylobacter jejuni* (most commonly from poultry but can be from pets) are some. Pets (e.g., cats and dogs) can be an occasional source of infection for other pathogens such as several tapeworms. Pet turtles are not as common as a source of *Salmonella* as in past years. Other pets (hamsters, rabbits, ferrets, etc.) can also be sources of some infections. This is a partial list.

Q: What are two of the most common causes of urethritis in males?

A: *Neisseria gonorrhoeae* and *Chlamydia trachomatis*. Gonococcal urethritis presents with a purulent discharge from the urethra, whereas nongonococcal urethritis is generally associated with a thinner, white mucus discharge. Treatment should cover both bacteria because there is a high incidence of coinfection. Ceftriaxone (especially), cefixime, cefotaxime, or penicillin G for *Neisseria gonorrhoeae* and doxycycline or azithromycin for *Chlamydia trachomatis* are the drugs of choice.

Q: What is the most common cause of UTIs?

A: *E. coli* (80–90%). *E. coli* is also the most common cause of pyelonephritis and pyelitis due to its ascension from the lower urinary tract. *Staphylococcus saprophyticus* accounts for 5–15% of UTI cases.

Q: What is the gold standard laboratory test for diagnosing a UTI?

A: Urine culture; in asymptomatic patients it requires at least 10^5 cfu/mL to be diagnostic.

Q: A 5-year-old child presents with sudden onset of gross hematuria, edema, hypertension, and renal insufficiency two weeks after a sore throat. What is the most likely diagnosis?

A: Acute poststreptococcal glomerulonephritis. Ninety-five percent of children completely recover within one month.

Q: What is the most common cause of acute renal failure in young children?

A: Hemolytic uremic syndrome

Q: Can tetanus occur after surgical procedures?

A: Yes. Most cases of tetanus in the United States occur after minor trauma, but there have been many reports of tetanus after general surgical procedures, especially those involving the GI tract. *Clostridium tetani* is part of the normal flora of the GI tract.

Q: What is the most common organism in wound infections?

A: *Staphylococcus aureus*. It is part of the normal skin flora.

Q: Most cases of pelvic inflammatory disease are caused by what two organisms?

A: *Neisseria gonorrhea* and *Chlamydia trachomatis*

Q: What are some of the most common organisms found in brain abscesses?

A: Enteric gram-negative bacilli and anaerobes, *Nocardia*, staphylococci, streptococci, and *Toxoplasma*

Q: Magnesium-ammonium phosphate stones most commonly are formed after what kind of infection?

A: UTIs due to *Proteus, Klebsiella, Pseudomonas, Providencia*, or any other urease-positive bacteria. The stones are also called infectious calculi, or struvite stones.

Q: What bacterial pathogens are associated with osteomyelitis in children?

A: *Staphylococcus aureus* is the most common. *Haemophilus influenzae* type b is most common in unimmunized children younger than 3 years old. There are others that are less common, but with test questions for any of these, obvious clues will be given. Depending on the reference, *Salmonella* may or may not be the most common in sickle-cell disease. Look at other information in the stem of the question to choose the correct answer.

Q: Hairy leukoplakia is due to what virus?

A: EBV. Hairy leukoplakia is usually found on the lateral aspect of the tongue in AIDS.

Q: A patient returns to the emergency room with fever, nausea, vomiting, and hypotension two days after having nasal packing placed for an anterior nosebleed. Toxic shock syndrome is considered. What organism causes toxic shock syndrome?

A: *Staphylococcus aureus* usually, but *Streptococcus pyogenes* can cause a similar disease.

Q: What are prions?

A: Infectious pathogens composed of proteins. They contain no nucleic acid (DNA or RNA). They appear to be modified proteins of normal cellular genes. The prion protein has the unique ability to affect the conformation of a normal protein (of unknown function) so that it becomes a prion. Thus, the only difference from normal protein is the conformation; they both (normal and prion) have the same amino acid sequences. Infections with prions are eventually fatal.

Q: What is the most common organism that causes pediatric acute otitis media?

A: *Streptococcus pneumoniae*, followed by *Haemophilus influenzae* and *Moraxella catarrhalis*

Q: What is the most common cause of epiglottitis?

A: *Haemophilus influenzae* type b. Most other upper respiratory tract infections are due to viruses (except for laryngotracheitis, which is also usually due to *H. influenzae* type b).

Q: What is the most common ocular infection in AIDS patients?

A: CMV retinitis occurs in 10–40% of AIDS patients and may lead to blindness.

Q: What organism causes river blindness?

A: Microfilariae of *Onchocerca volvulus* (a nematode) cause this disease, which is also known as onchocerciasis. River blindness is a major cause of blindness worldwide. It is characterized by uveitis, chorioretinal changes, and optic atrophy.

Q: How is chlamydial inclusion conjunctivitis transmitted?

A: Contact. Neonates can acquire the organism as they pass through the birth canal. Others can acquire it via contaminated swimming pool water, towels, hands, and so on. It can be spread from person-to-person by sexual contact.

Q: What is the most common cause of ophthalmia neonatorum in the United States?

A: *Chlamydia trachomatis*. *Neisseria gonorrhoeae* is another cause.

Q: Why is Rh status important in a pregnant patient?

A: If the mother is Rh negative and the fetus is Rh positive, erythroblastosis fetalis, also known as hydrops fetalis or hemolytic disease, can result. If the maternal blood and that of the baby become mixed at the time of birth (or earlier), any future babies would be at risk because the mother's body would have produced antibodies to destroy the blood of future babies. Rh immunoglobulin should be given to all Rh-negative mothers at about 28 weeks gestation and again within 72 hours of giving birth.

Q: What is the most common cause of vaginitis?

A: *Candida albicans*

Q: What is kala azar?

A: Kala azar is also called disseminated visceral leishmaniasis. It is caused by *Leishmania donovani*.

Q: A patient presents with a two-day history of vaginal itching and burning. On exam, you note a thin yellowish-green bubbly discharge and petechiae on the cervix (also known as "strawberry cervix"). What lab test do you perform and what results do you expect to find?

A: Look at the discharge mixed with saline under a microscope. If you see a motile pear-shaped protozoan (*Trichomonas vaginalis*) with flagella, then the patient should be treated with metronidazole.

Q: A 16-year-old sexually active girl is complaining of a heavy but thin vaginal discharge with an unpleasant odor. When you add 10% KOH to the discharge, the combination emits a fishy odor. What would you expect to see on microscopic examination?

A: "Clue cells." These are epithelial cells with bacilli attached to their surfaces. This patient has vaginosis, a mixed infection caused by loss of major *Lactobacillus* normal flora species. The primary pathogens are *Gardnerella vaginalis*, *Mobiluncus* + 9 to 17 others (mostly anaerobes). The patient should be treated with metronidazole.

Q: What infection does *Haemophilus ducreyi* cause?

A: Chancroid (soft chancre; genital ulcers). It is an STD more common in Africa and Asia than in the United States or Europe.

Q: What do the terms "condyloma acuminatum" and "condyloma latum" refer to and what is their cause?

A: They are genital warts due to human papillomaviruses (especially HPV-6 and HPV-11) and *Treponema pallidum*, respectively.

Q: What is the most common cause of septic arthritis in young adults?

A: Disseminated gonococcal infection (*Neisseria gonorrhoeae*)

Q: What antimicrobials should be given to a patient with gonorrhea?

A: Ceftriaxone (drug of choice) or cefixime, cefotaxime, or penicillin G for *Neisseria gonorrhoeae* and doxycycline or azithromycin for probable chlamydia. Either of the latter is given because of the high incidence of coinfection with *Chlamydia trachomatis*.

Q: Is there a Lyme disease vaccine?

A: Not any more. It was pulled from the market in 2002.

Q: What is the predominant bacterium in a healthy female's vagina?

A: *Lactobacillus*. These bacteria are also responsible for the low vaginal pH (due to lactic acid production) that is an important detriment to colonization by potential pathogenic bacteria. A decrease in the *Lactobacillus* population results in a higher pH, allowing for other bacteria to proliferate. This can result in vaginal infections such as bacterial vaginosis. Normal vaginal strains of lactobacilli also produce hydrogen peroxide that protects against colonization by other bacteria. The lactobacilli also may protect against HIV infection by significantly reducing the HIV load in the vaginal tract.

Q: A 16-year-old boy presents to your office with a maculopapular rash on his palms and soles. He complains of headaches and general weakness. On exam, you find he has multiple condyloma latum and lymphadenopathy. What is the diagnosis?

A: Secondary syphilis. This develops 6–9 weeks after the development of the syphilitic chancre, which resolves by this time. If it is allowed to progress untreated, tertiary syphilis may develop years later. This can affect all the tissues in the body, including those of the CNS and the heart. Treatment of secondary syphilis is penicillin G.

Q: What causes toxic shock syndrome?

A: *Staphylococcus aureus*, due to toxic shock syndrome toxin-1, which is a superantigen. Other organisms causing toxic shock syndrome include *Streptococcus pyogenes* and less commonly *Pseudomonas aeruginosa* and *Streptococcus pneumoniae*. Tampons, intrauterine devices, septic abortions, contraceptive sponges, soft tissue abscesses, osteomyelitis, nasal packing, and postpartum infections can be predisposing factors.

Q: What dermatological changes occur with toxic shock syndrome?

A: Initially, the patient has a blanching erythematous rash that lasts for three days. Ten days after the start of the infection there is a desquamation on the palms and soles.

Q: A 16-year-old girl presents with fever, aches, and painful genital sores. Two days ago they looked like blisters until they popped and became painful. What would you expect to find on culture of the vesicular fluid?

A: No bacteria, but virus culture would most likely grow herpes simplex virus, probably type 2 because it is more common than type 1. A Tzanck smear prepared from the lesion and stained by Giemsa or Papanicolaou method will show multinucleated giant cells.

Q: What subtypes of HPV are associated with cervical cancer?

A: HPV types 16 and 18 are the most common risk factors for cervical dysplasia that can lead to cervical cancer. Multiple sex partners and early onset of sexual activity are also risk factors for cervical cancer. The use of the HPV vaccine in females should result in fewer cases.

Q: Name some viruses that cause pneumonia.

A: Parainfluenza viruses (including human metapneumovirus), influenza virus, adenovirus, and RSV

Q: What is the most common cause of vaginitis in girls who are pubertal but not sexually active?

A: *Candida albicans*

Q: What are the most common causes of acute vaginitis?

A: *C. albicans*, and *Trichomonas vaginalis*

Q: Besides *Gardnerella vaginalis*, what other bacteria are involved in bacterial vaginosis?

A: Bacterial vaginosis is a mixed infection involving *Mobiluncus* plus 9 to 17 others, mostly anaerobes.

Q: How is bacterial vaginosis diagnosed?

A: Microscopic observation of clue cells, pH > 4.5, and a positive whiff test

Q: What is the treatment for bacterial vaginosis?

A: Metronidazole (clindamycin if pregnant)

Q: How is the laboratory diagnosis of vaginitis due to *Trichomonas vaginalis* made?

A: Diagnosis is usually made by examination of a wet mount preparation of the vaginal discharge. The movement of the flagellated protozoa can be seen.

Q: How does an adolescent girl acquire vaginitis due to *Trichomonas vaginalis*?

A: Vaginitis due to *T. vaginalis* is primarily an STD.

Q: What is the treatment for vaginitis due to *T. vaginalis*?

A: Metronidazole, which can be given orally as a single dose

Q: A 9-week-old female infant has had a white malodorous discharge from her vagina for more than seven weeks. Her physical examination is normal except for the aforementioned discharge. A wet prep of the discharge shows motile trichomonads. How did the infant acquire this infection?

A: A small number of vaginally delivered female infants may acquire vaginitis due to *T. vaginalis* from their untreated mothers during delivery.

Q: What is the purpose of the RPR test?

A: The RPR test detects antibodies formed against epitopes on the surface of *Treponema pallidum* that cross-react with a cardiolipin–cholesterol–lecithin complex. It is a nonspecific nontreponemal test for syphilis that does not directly detect *T. pallidum*. You will not need to know what RPR stands for, but you should know how the test works and that this is a presumptive test for syphilis. VDRL is a similar nontreponemal presumptive test. Classically, the RPR and VDRL tests were presumptive only if there were no other lab values and confirmatory tests were needed. However, a recombinant antigen EIA is available as a screening procedure for detection of syphilis specific antibodies. For a positive EIA screen, the use of the RPR test in combination with RPR quantitation can be used as the confirmatory method. This approach reduces the total cost of the evaluation because it reduces the use of the more expensive FTA testing in most cases where the EIA and RPR are in agreement.

Q: Name some causes of a false-positive RPR?

A: Pregnancy, infection (e.g., infectious mononucleosis, tuberculosis, and endocarditis), lupus, recent immunization, intravenous drug abuse, and rheumatoid arthritis are some. False positives are due to the presence of anticardiolipin antibodies.

Q: What is the most prevalent STD among adolescent girls?

A: HPV. A vaccine was approved in the United States by the Food and Drug Administration (FDA) in 2006 for HPV types 6, 11, 16, and 18. The risk of male-to-female transmission of HPV infection can be reduced with male condom use.

Q: What is the most common cause of abnormal vaginal discharge in sexually active adolescents?

A: *Trichomonas vaginalis*

Q: Cervical cancer is often associated with what STD?

A: HPV (especially types 16 and 18), the cause of genital warts (condyloma acuminata)

Q: What is the most common etiology of genital ulcers in the United States?

A: Herpes simplex, type 2

Q: How can one easily make the diagnosis of herpes simplex?

A: By Tzanck smear, which shows multinucleated giant cells, or by viral culture

Q: At what point should antiretroviral treatment begin in an asymptomatic patient with HIV?

A: When CD4+ T cells fall below 350 cells/mm^3.

Q: What is the drug of choice for *Pneumocystis* pneumonia for patients with HIV infection?

A: Trimethoprim-sulfamethoxazole

Q: When should prophylactic therapy begin for *Pneumocystis* pneumonia for patients with HIV infection?

A: When they have a CD4+ T-lymphocyte count of less than 200/μL or a history of oropharyngeal candidiasis.

Q: In what age group is poststreptococcal glomerulonephritis most common?

A: The peak incidence is between ages 3 and 7, but you only need to know that it is most common in children.

Q: A patient who has lived his entire life in Omaha, Nebraska is diagnosed with typhoid fever. What question should be asked of this patient?

A: "Have you recently traveled out of the country?" Most cases in the United States are contracted this way.

Q: What is used to control outbreaks of meningococcal meningitis?

A: Rifampin and ceftriaxone are used as chemoprophylaxis for contacts. A vaccine for *Neisseria meningitidis* groups A, C, Y, and W-135 is available and widely used even though most outbreaks are caused by strains A, B, C, and W-135. The vaccine is recommended for freshmen college students living in dormitories.

Q: Antibiotic therapy may result in pseudomembranous colitis in an individual being treated for another infection. How?

A: Suppression of normal intestinal flora with broad-spectrum antibiotics allows for the overgrowth of *Clostridium difficile*.

Q: A teenage girl is going on a summer trip to Benin, on the west coast of Africa. What immunizations and prophylactic treatments must she receive before departing?

A: Hepatitis A vaccine, oral polio vaccine, DTaP vaccine, live oral typhoid vaccine, measles vaccine, yellow fever vaccine, and mefloquine prophylaxis for malaria

Q: Why is it important to identify food service workers with furunculosis?

A: Furunculosis is most commonly caused by *Staphylococcus aureus*. Staphylococcal enterotoxin is a leading cause of food poisoning.

Q: Which has a longer incubation period in food poisoning: *Staphylococcus* or *Salmonella*?

A: *Salmonella*. It is generally ingested in small doses and then multiplies in the GI tract. Symptoms occur 6–48 hours after ingestion. With *Staphylococcus*, the disease process is dependent not on the reproduction of the organism within the host but rather on the enterotoxin it has already produced in the contaminated food.

Q: A 27-year-old man stepped on a rusty nail. He cannot remember the last time he got a tetanus shot. What should you do in regards to his immunization?

A: Treat with a tetanus immunoglobulin intramuscularly and tetanus toxoid (Td). Note: With respect to treatment, there is nothing special about rusty nails and tetanus; any puncture wound can drive *C. tetani* spores into the subcutaneous tissues.

Q: At what age is the peak incidence of *Haemophilus influenzae* meningitis?

A: Seven to 12 months.

Q: Aside from those who are immunocompromised, what patients should *not* receive live virus vaccines?

A: Pregnant women should not receive live viruses, especially MMR (due to the tetragonic potential). Oral polio should be avoided in anyone in close contact with an immunocompromised person because of the virus's ability to spread.

Q: How many millimeters in diameter indicate a positive reaction to the Mantoux skin test in a person with HIV?

A: Five millimeters or greater. In most healthy individuals, induration must be equal to or greater than 10 mm if they have risk factors for TB. For those with no risk factors, induration must be equal to or greater than 15 mm to be positive.

Q: When should the influenza vaccine be given?

A: In September or October, about 1–2 months before flu season begins, but as late as January. If formulated correctly (i.e., the correct serotypes of virus are included), the vaccine is protective against influenza A and B.

Q: The larvae of what parasite can penetrate and infect muscle tissue after ingestion of raw or undercooked infected meat?

A: *Trichinella spiralis*, the cause of trichinosis, aka trichinellosis. The disease, once very common in the United States after eating pork, is not as common as it once was (currently about 10–12 cases annually). Persons at highest risk are those eating raw or undercooked meat of wild animals, such as bears.

Q: Amantadine can be used for therapy of which influenza virus?

A: Historically, influenza A only. Unfortunately, because of resistance, the drug is no longer recommended in the United States.

Q: What is the prophylactic drug of choice for contacts of a child with meningococcal meningitis?

A: Rifampin

Q: What is an alternative name for the smallpox virus?

A: Variola virus. There are two variants: variola major (causes the most serious disease with a higher mortality rate) and variola minor (causes a milder form of disease).

Q: What is the principle vector for malaria?

A: *Anopheles* mosquitoes

Q: What is the principle vector for yellow fever and dengue fever?

A: *Aedes aegypti* mosquitoes

Q: What is the principal route of transmission of schistosomiasis?

A: Skin contact with water from lakes or rivers that are infested with the cercariae (larva) of the schistosomes. The cercariae penetrate the skin and travel through the heart, lungs, and circulatory system until they reach the portal veins where they develop into nonpathogenic worms.

Q: What is the difference between the classic and alternate complement cascades?

A: You do not really need to know all the differences. It is more important to know what initiates the two and what the result is. The classic complement cascade is initiated primarily by an antigen–antibody complex and the alternate pathway is initiated by pathogens (e.g., bacteria, fungi, etc.) or their products (e.g., LPS). IgG (not IgG$_4$) and IgM are the primary immunoglobulins capable of fixing complement. In the classic pathway, C1 (actually C1q) binds to the Fc portion of the H chains to begin the cascade (binding of C1q requires two IgGs or one IgM). The cascade involves many other components, including C1r, C1s, and C2 through C9 and their cleavage products (e.g., C2a, C2b, C4a, C4b, etc.). C1, C2, C3, C4, and above are involved in the classic pathway. The alternate pathway begins with C3 but does not involve C1, C2, or C4. Other components involved in the alternate pathway are factors B and D, and properdin. The final stages of the pathways result in the formation of a membrane attack complex that can cause cell lysis. Activation of complement can lead to cell lysis, opsonization, promotion of inflammation, chemotaxis of polymorphonuclear neutrophilic leukocytes, and so on. It might be worthwhile to review the complement cascades in a textbook.

Q: What is being measured in a complement fixation test?

A: Either antibody or antigen, whichever is unknown. In the complement fixation test, a complement-fixing antibody is allowed to react with an antigen in the presence of complement. In the process, complement components are used up. If sensitized erythrocytes (anti-red blood cell + red blood cells) are added, no lysis occurs because the complement components have been depleted. In the presence of complement, the sensitized red blood cells lyse.

Q: What is the most important factor for reducing infectious disease transmission in daycare centers?

A: Hand washing

Q: A 4-year-old child who attends day care has impetigo. When should the child be allowed to return to day care?

A: Twenty-four hours after treatment has been initiated

Q: What are the major sequelae of untreated streptococcal infections?

A: Acute glomerulonephritis and rheumatic heart disease

Q: What parasite is most commonly associated with CNS involvement in an AIDS patient?

A: *Toxoplasma gondii*

Q: What organisms most commonly cause bacterial conjunctivitis?

A: *Haemophilus influenzae* and *Streptococcus pneumoniae* especially, but also *Neisseria gonorrhoeae* and *Chlamydia trachomatis*

Q: Describe the presentation of staphylococcal scalded skin syndrome.

A: Staphylococcal scalded skin syndrome is usually contracted by children younger than age 5 years and results in an intraepidermal cleavage beneath the stratum granulosum. The disease begins after an upper respiratory tract infection or purulent conjunctivitis. Initially, the lesions are tender, erythematous, and scarlatiniform. They are usually found on the face, neck, axillae, and groin. Later, the skin peels off in sheets with lateral pressure, and a positive Nikolsky sign (gentle rubbing of the skin will cause it to slough away) is displayed.

Q: What are the two most common causes of fatal anaphylaxis?

A: Drug reactions (especially penicillin) and stings (bees and wasps)

Q: Which type of hypersensitivity reaction is responsible for anaphylaxis?

A: Type I (IgE mediated). The four hypersensitivities are as follows:

- Type 1: immediate; IgE binds allergen; includes mast cells and basophils (e.g., food allergy, asthma)
- Type 2: cytotoxic; IgG and IgM antibody reacts with antigen on cell surface resulting in complement activation (e.g., blood transfusion reaction)
- Type 3: immune complex; deposition of antigen–antibody complexes in tissues (e.g., Arthus reaction, poststreptococcal glomerulonephritis)
- Type 4: delayed; cell mediated (e.g., contact hypersensitivity [poison ivy], skin tests [tuberculin])

Q: Describe food-borne botulism.

A: Botulism exotoxin is secreted by *Clostridium botulinum*. It affects the myoneural junction and prevents the release of acetylcholine. In the United States it is usually caused by the ingestion of foods that were inadequately prepared (not sterilized). The most common neurological complaints are related to the bulbar musculature. Neurological symptoms usually occur within 24–48 hours of ingestion of the contaminated food. Symmetrical descending paralysis muscle paralysis and weakness typically spread rapidly to involve all muscles of the trunk and extremities.

Q: What are the most common initial symptoms in botulism poisoning?

A: Visual disturbances, headache, dizziness, weakness, malaise, and a dry mouth. Nausea and vomiting are found in about 35% of the cases. Symptoms usually appear within 24–72 hours after exposure.

Q: Describe the rash that is present with roseola (exanthem subitum).

A: The rash is usually found on the trunk and the neck and is maculopapular. Roseola is due to HHV-6.

Q: What are the two most common causes of acute viral diarrhea?

A: Norovirus (adults and older children) and rotavirus (children under 2 and infants)

Q: What is the most common food-borne pathogen?

A: *Staphylococcus aureus*. This pathogen produces vomiting within a short period of time (four hours average) after ingestion of the contaminated food containing the preformed enterotoxin.

Q: What bacterial gastroenteritis is associated with the consumption of seafood?

A: Consumption of improperly prepared oysters, crabs, shrimp, and so on can lead to gastroenteritis due to *Vibrio parahaemolyticus*. It is relatively common in some countries such as Japan. Wound infection by this organism is rare but has a high mortality rate.

Q: What are the protozoa that are commonly associated with diarrhea?

A: *Giardia duodenalis, Entamoeba histolytica, Cryptosporidium, Cyclospora,* and *Balantidium coli*

Q: What protozoa cause bloody diarrhea?

A: *E. histolytica* and *B. coli*. A watery diarrhea is more characteristic of *G. duodenalis* (rarely bloody), *Cyclospora,* and *Cryptosporidium*. On most standardized exams, you probably will not need to know about any other protozoa that might cause diarrhea.

Q: What is the most common parasitic cause of diarrhea in the United States?

A: *Giardia duodenalis* (previously *G. lamblia*). Infected individuals present with foul-smelling floating stools, abdominal pain, and profuse watery diarrhea. The parasite can be identified by microscopic observation of trophozoites.

Q: What causes the diarrhea commonly seen in AIDS patients?

A: *Cryptosporidium* is most common. A positive acid-fast stain is important in the diagnosis of this parasite. Patients present with profuse watery diarrhea that is not bloody.

Q: What is the most common cause of impetigo?

A: *Staphylococcus aureus* or *Streptococcus pyogenes* (group A), depending on what you read

Q: What is the most common cause of bullous impetigo?

A: *Staphylococcus aureus*

Q: Name some diseases in which the clinical manifestations are due to a superantigen?

A: Staphylococcal and streptococcal toxic shock syndromes are the most intensely studied. Others include some of the enterotoxins involved in staphylococcal food poisoning, exfoliatin of staphylococcal scalded skin syndrome, the rash of scarlet fever, and possibly HIV-1 and other viruses. As research continues, more examples are emerging.

Q: What is the receptor site for superantigens?

A: They bind to the outside of a class II MHC molecule on an antigen-presenting cell and a T-cell receptor (the <u>v</u>ariable portion of the <u>beta</u> chain [Vβ]) on a T-helper cell).

Q: What happens after superantigens bind to class II MHC molecules and T-cell receptors?

A: There is a nonspecific activation of T-helper cells resulting in the release of large amounts of cytokines such as IL-1, IL-2, and TNF.

Q: What group of anaerobes is consistently beta-lactamase positive?

A: The *Bacteroides fragilis* group. Therefore the penicillins and cephalosporins are not drugs of choice.

Q: What are the drugs of choice for the *B. fragilis* group?

A: Metronidazole or clindamycin

Q: What is the most common cause of viral pneumonia in the pediatric patient?

A: RSV

Q: Name two viral illnesses that can lead to Reye syndrome.

A: Chickenpox (varicella; varicella-zoster virus) and influenza (influenza B and A). Aspirin should be avoided in children.

Q: What is the most common cause of a UTI in a female child?

A: *E. coli* is always the right answer for UTIs unless there is additional information (e.g., urease positive = *Proteus* or *Klebsiella*) that points you in another direction.

Q: List some infections of *Yersinia* and the species involved.

A: Plague, *Yersinia pestis*; gastroenteritis, *Y. enterocolitica* (children) and *Y. pseudotuberculosis* (appendicitis-like symptoms more often than diarrhea)

Q: What is the appropriate treatment for a nonimmunized individual exposed to hepatitis B?

A: Treat the individual with hepatitis B immune globulin and consider vaccination.

Q: What is the nature of the lesions of tertiary syphilis?

A: They are noninfectious granulomas.

Q: What are the granulomatous lesions of tertiary syphilis called?

A: Gummas

Q: What drugs can be used for the treatment of chronic hepatitis B?

A: Adefovir, interferon alfa-2b, pegylated interferon alfa-2a, lamivudine, entecavir, and telbivudine

Q: A researcher using an HIV-susceptible cell line developed a subclone of the cells that did not possess the CCR5 receptor. What would be the effect on the subclone with respect to HIV susceptibility?

A: The subclone would be resistant to infection because the CCR5 chemokine receptor is a required coreceptor for HIV infection.

Q: What infectious agents commonly cause laryngotracheobronchitis (croup)?

A: Parainfluenza viruses, especially types 1 and 2, but type 3 can also be a cause. The disease presents as a barking cough and is most common in the fall. Symptoms typically last about two days.

Q: What gynecological infection presents with a malodorous, itchy, white to grayish and sometimes frothy vaginal discharge?

A: Trichomoniasis

Q: Which streptococcus causes acute poststreptococcal glomerulonephritis?

A: *Streptococcus pyogenes* (group A beta-hemolytic streptococcus)

Q: Describe the symptoms and signs of varicella (chickenpox).

A: The onset of varicella rash is 1–2 days after prodromal symptoms of slight malaise, anorexia, and fever. The rash begins on the trunk and scalp, appearing as faint macules and later becoming vesicles. As opposed to smallpox, chickenpox lesions typically are in all stages of development.

Q: Describe the signs of roseola (aka roseola infantum, exanthema subitum, sixth disease).

A: Roseola usually affects children ages 6 months to 3 years. It is characterized by high fever (103–105°F) that begins abruptly and lasts 3–5 days. A rash appears as the temperature drops to normal.

Q: What causes roseola?

A: Primarily, HHV-6. There are two genetic variants, HHV-6A (no known disease associations) and HHV-6B (a mononucleosis syndrome, encephalitis, pneumonitis, others). Recently, strong evidence has shown that one particular type of epilepsy (mesial temporal lobe epilepsy) is associated with HHV-6B infection.

Q: How does a child with erythema infectiosum (fifth disease) present?

A: Fifth disease typically does not infect infants or adults. There are no prodromal symptoms. The illness usually begins with the sudden appearance of erythema of the cheeks (slapped cheek syndrome), followed by a maculopapular rash on the trunk and extremities that evolve into a lacy pattern.

Q: What causes fifth disease?

A: Parvovirus B16

Q: What organism most commonly causes a septic joint in a child?

A: *Staphylococcus aureus*

Q: What organism most commonly causes a septic joint in an adolescent?

A: *Neisseria gonorrhoeae*

Q: What organism most commonly causes bacterial pneumonia in adults?

A: *Streptococcus pneumoniae*

Q: What triad is associated with Reiter syndrome (reactive arthritis)?

A: Nongonococcal urethritis, polyarthritis, and conjunctivitis. The conjunctivitis is the least common and occurs in only 30% of the patients. Acute attacks respond well to nonsteroidal anti-inflammatory drugs.

Q: What are the complications of streptococcal impetigo?

A: Streptococcal-induced impetigo (*Streptococcus pyogenes*) can result in poststreptococcal glomerulonephritis. However, it has not been shown to be associated with rheumatic fever.

Q: Describe the key features of the Rocky Mountain spotted fever rash.

A: Rocky Mountain spotted fever is caused by *Rickettsia rickettsii* and is transmitted by ticks. Patients become sick with a high fever, headache, chills, and muscular pain. Around the fourth day of fever, a rash begins on the wrists and ankles that spreads centripetally (toward the trunk).

Q: How does an infant with botulism present?

A: Infant botulism presents with lethargy and failure to thrive. Most infants survive with no significant sequelae. In rare instances paralysis and death can occur. Raw honey has been considered a common source of *Clostridium botulinum* spores.

Q: Does food-borne botulism produce fever?

A: No. This can be important in differentiating neurological symptoms in a sick patient who could have diphtheria. Both botulism and diphtheria are more common in developing countries.

Q: Describe a patient with chlamydial pneumonia.

A: Pneumonia due to *Chlamydophila* (*Chlamydia*) *pneumoniae* presents as a mild "walking pneumonia" similar to that caused by *Mycoplasma pneumoniae*. Chronic cough is prominent and the patient is afebrile.

Q: What is the drug treatment for *Giardia duodenalis*?

A: Metronidazole is most commonly used in the United States for giardiasis, but tinidazole is also effective.

Q: In what types of patients is staphylococcal pneumonia likely?

A: Patients that are hospitalized, debilitated, or abusing drugs. *Staphylococcus aureus* is usually a cause of pneumonia in the immunocompromised or after other pulmonary disease. Staphylococcal pneumonia secondary to influenza infection can be particularly severe.

Q: What is the most common bacterium causing septic arthritis?

A: It depends. *Staphylococcus aureus* is most common in children. *Neisseria gonorrhoeae* is most common in sexually active adults.

Q: What causes swimmer's itch?

A: Cercariae of some schistosomes that normally infect animals such as birds. After invasion of the human skin, the infection remains localized. This usually mild self-limiting disease is most common in developing countries but is found in most developed countries as well, including the United States. The disease is also called cercarial dermatitis.

Q: What is the vector of trypanosomiasis?

A: Tsetse fly (Africa)

Q: What is the presentation of a patient with respiratory diphtheria?

A: Sore throat, dysphagia, fever, and tachycardia. A "dirty," tough, gray fibrinous pseudomembrane may be present in the oropharynx. It may be so firmly attached to the mucosa that its removal causes profuse bleeding. *Corynebacterium diphtheriae* exotoxin acts directly on cardiac, renal, and nervous systems. It causes ADP-ribosylation of elongation factor 2, thereby inhibiting protein synthesis.

Q: When does an elevation of HBsAg occur in relation to symptoms of hepatitis B?

A: HBsAg rises before clinical symptoms of hepatitis B.

Q: Is hepatitis A associated with jaundice?

A: It depends on age. Jaundice can occur at any age but occurs more frequently in adults (60–70%) than in children (10–20%).

Q: Should salmonellosis be treated with antimicrobials?

A: Only if symptomatic infection persists or is severe. Most cases resolve without treatment. Use of antimicrobials may lead to a carrier state and can select for antibiotic-resistant strains.

Q: What is the most common cause of acute food-borne disease?

A: Probably *Staphylococcus aureus* and the enterotoxin it secretes in food

Q: What are the common features of *Vibrio parahaemolyticus* infection?

A: This organism causes food poisoning associated with ingestion of oysters, clams, and crabs. Symptoms include cramps, vomiting, dysentery, and explosive diarrhea. It can also be a cause of seawater-associated wound infections. After the Katrina hurricane of 2005, two of three infected individuals died.

Q: What organisms are most commonly present in pulmonary abscesses such as aspiration pneumonia?

A: Mixed anaerobes from the oral cavity, but sometimes aspirated bacteria from the stomach

Q: What is the most common cause of food-borne viral gastroenteritis?

A: Norovirus causes more than half of all viral outbreaks of gastroenteritis in adults. Rotavirus is most common in infants and young children.

Q: What is the most common cause of epiglottitis?

A: *Haemophilus influenzae* historically has been the most common, especially among children around age 5. Because of the use of the Hib vaccine, the incidence is dropping.

Q: What is the common name for *Dermacentor andersoni*? What are the disease associations?

A: Wood tick. It is a vector for Rocky Mountain spotted fever and Colorado tick fever. Rocky Mountain spotted fever is also carried by *Dermacentor variabilis* (dog tick) and *Amblyomma americanum* (lone star tick).

Q: What is the most common site of infectious arthritis?

A: The knee, followed by the hip. *Staphylococcus aureus* is the most common cause.

Q: A positive TB test is what type of hypersensitivity reaction?

A: Type 4, cell-mediated delayed hypersensitivity; neither complement nor antibodies are involved

Q: What organism is most commonly found in cutaneous abscesses?

A: *Staphylococcus aureus*

Q: Describe the Gram stain appearance of *S. aureus*.

A: Gram-positive cocci in grape-like clusters

Q: What is the most common cause of cellulitis in children younger than 3 years old?

A: *Haemophilus influenzae* type b, but as with other *H. influenzae* infections, the use of the Hib vaccine has resulted in fewer infections than before the vaccine.

Q: *Clostridium tetani* produces a neurotoxin. Is it an endotoxin or an exotoxin?

A: The tetanus toxin, tetanospasmin, is an exotoxin. Because *C. tetani* is gram-positive, it cannot have an endotoxin (i.e., it has no LPS).

Q: What parasite causes a greenish-gray frothy vaginal discharge with mild itching?

A: *Trichomonas vaginalis*. About 20% of the time the cervix has a strawberry appearance. The infection can be confirmed with a wet mount.

Q: A 4-year-old presents with irritability, weakness, lethargy, dehydration, edema, petechiae, and hepatosplenomegaly 10 days after an episode of gastroenteritis. Laboratory results reveal a low platelet count and hemoglobulin level, what is the most likely diagnosis?

A: Hemolytic uremic syndrome

Q: What is the diagnostic triad for hemolytic uremic syndrome?

A: Microangiopathic anemia, acute renal failure, and thrombocytopenia

Q: What causes chancroid?

A: *Haemophilus ducreyi*. The resulting ulcer is painful; the chancre of syphilis is painless.

Q: What is the most common infectious cause of UTIs in females?

A: *E. coli*

Q: When does erythema migrans appear in Lyme disease?

A: In stage I: 3–30 days after the tick bite. Its appearance is diagnostic, but commonly is absent.

Q: What is the vector and causative organism of Lyme disease?

A: The vector is the tick *Ixodes scapularis* and the organism is the spirochete *Borrelia burgdorferi*. It is the most frequently transmitted tick-borne disease in the U.S.

Q: What causes Q fever?

A: *Coxiella burnetii*, which is a rickettsia-like organism

Q: A 10-year-old boy with a history of ventricular septal defect is about to have an impacted tooth removed. For what infection is he at risk?

A: Endocarditis due especially to viridans streptococci such as *Streptococcus mutans* and *S. salivarius*. Viridans streptococci are found in the oral cavity and stick easily to damaged heart valves. There is no such organism as *Streptococcus viridans*.

Q: Name some enterotoxin-producing organisms that can cause food poisoning.

A: *Clostridium perfringens*, *Staphylococcus aureus*, *Vibrio cholerae*, *Bacillus cereus*, and *E. coli*. There are others such as *Campylobacter jejuni* and *Yersinia enterocolitica* (children), but you may not need to know about them because their role in disease is still being investigated.

Q: What is the most common cause of nosocomial bacteremia?

A: Coagulase-negative staphylococci, such as *S. epidermidis* and *S. saprophyticus*. *Staphylococcus aureus* is the second most common.

Q: A patient developed a hand infection after striking another person in the mouth. In choosing empiric therapy, what organisms should be considered?

A: This is a case of clenched-fist injury. Classically, these injuries can involve *Eikenella corrodens*, but other organisms, including *Staphylococcus aureus* and oral anaerobes, should also be considered.

Q: What would you expect to find in the hippocampus of a patient with rabies?

A: Negri bodies. The incubation period for rabies is 30–60 days. Treatment must begin before the appearance of symptoms and includes cleaning of the wound, rabies immune globulin, and rabies vaccine. As of this writing, there are only six known survivors of rabies after symptoms began.

Q: A patient presents with mucocutaneous candidiasis. Laboratory findings include low calcium and T cells. Immunoglobulin levels are normal. This patient probably has what syndrome?

A: DiGeorge syndrome. Besides hypoplasia of the parathyroid and thyroid glands, you may also find any of the following: right-sided aortic arch, esophageal atresia (closure), bifid uvula, atrial and ventricular septal defects, short philtrum of upper lip, hypertelorism (eyes wider apart), epicanthal folds, mandibular hypoplasia, and low-set notched ears.

Q: What organism causes Weil disease?

A: *Leptospira interrogans*. Humans frequently are infected by direct contact with animal urine, such as from dogs. Direct contact includes eating food or drinking water (especially ingestion while swimming) that has been contaminated. Usually, the term Weil disease is applied for severe manifestations of leptospirosis.

Q: Prior infection with what organism(s) sometimes precedes Guillain-Barré syndrome?

A: Due to an autoimmune response to the myelin sheath of nerves, most cases have been associated with infections, especially *Campylobacter jejuni* (about half of all cases of Guillain-Barré syndrome) and other organisms such as herpesvirus, HIV, EBV (infectious mononucleosis), and *Mycoplasma pneumoniae*. Other factors that may lead to Guillain-Barré syndrome are vaccination, surgery, lymphoma, hepatitis, or systemic lupus erythematosus.

Q: The addition of KOH to vaginal secretions resulted in a fishy odor. What is the infection and what microscopic test can be done for confirmation?

A: The microscopic observation of clue cells indicates bacterial vaginosis caused by *Gardnerella vaginalis* and other normal flora organisms. It is a mixed infection, but *G. vaginalis* is a major pathogen.

Q: A vaginal discharge with the consistency of cottage cheese suggests infection with what organism?

A: *Candida albicans*

Q: The number of epithelial cells seen on microscopic examination of sputum smears is indicative of the presence or absence of significant contamination of the specimen with oral secretions. How many squamous epithelial cells per low power field (lpf; 100×, total magnification) are considered indicative of such contamination?

A: 10 or more

Q: How many of the more than 80 antigenic capsular types of *Streptococcus pneumoniae* are included in the current adult pneumococcal vaccine?

A: Although there are only 23 capsular serotypes in the vaccine, it covers most strains usually found in infections.

Q: At what age should the pneumococcal vaccine (PPV23) be administered among healthy adults?

A: Age 65; earlier for those with chronic illnesses (diabetes, alcoholism, liver disease, etc.)

Q: What category of bacteria is most important in community-acquired aspiration pneumonia?

A: Anaerobic bacteria

Q: What community-acquired pneumonia is typically spread by inhalation of microdroplets from common water supplies?

A: Legionnaires' disease. There is no person-to-person transfer.

Q: What is the most common etiological agent of atypical pneumonia?

A: *Mycoplasma pneumoniae*. *Chlamydophila pneumoniae* is probably the second most common.

Q: What is the most common cause of community-acquired bacterial pneumonia among patients infected with HIV?

A: *Streptococcus pneumoniae*. They are also at increased risk for pneumonia due to other organisms such as *Haemophilus influenzae* and *Pseudomonas aeruginosa*.

Q: What is the single most important risk factor for hospital-acquired bacterial pneumonia?

A: Endotracheal intubation

Q: What are important risk factors for the development of anaerobic lung abscess?

A: Poor oral hygiene and a predisposition toward aspiration

Q: Splinter hemorrhages, Osler nodes, Janeway lesions, and petechiae can be indications of what disease process?

A: They are physical signs associated with infective endocarditis. Osler nodes are usually nodular and painful. In contrast, the macular Janeway lesions are painless.

Q: In what other conditions besides infective endocarditis can Osler nodes be found?

A: Nonbacterial thrombotic endocarditis, gonococcal infections, and hemolytic anemia

Q: What is the most common pathogen isolated in children with bronchiolitis?

A: RSV, especially in infants younger than 1 year old

Q: What specimen should be sent to the bacteriology laboratory for diagnosis of bacterial endocarditis?

A: Blood for culture

Q: Acute renal failure, hemolytic anemia, and thrombocytopenia comprise the triad for which syndrome?

A: Hemolytic uremic syndrome

Q: What is the most common etiologic agent isolated from blood cultures taken from children with epiglottitis?

A: *Haemophilus influenzae* type b (Hib). Bacteremia is common in epiglottitis with this organism. Airway obstruction can result in death. The incidence has decreased since the introduction of the Hib vaccine.

Q: What group of organisms is the most common cause of croup?

A: Viruses, especially parainfluenza virus, but also adenoviruses, influenza virus, and a few others

Q: A 6-year old boy complains of dysphagia and fever. Tonsillar exudates and anterior cervical adenitis are noted. What sign differentiates *Streptococcus pyogenes* from other causes of pharyngitis?

A: A sandpaper-like rash spreading from the chest down to the trunk and to the extremities often appears by the second day after *S. pyogenes* pharyngitis.

Q: What are the most common causes of bacterial conjunctivitis?

A: *Haemophilus influenzae* and *Streptococcus pneumoniae* are the most common. Chlamydial (inclusion) conjunctivitis (*Chlamydia trachomatis*) commonly affects sexually active teens and young adults and is the most frequent infectious cause of neonatal conjunctivitis in the United States. *Neisseria gonorrhea* can also cause a sexually transmitted conjunctivitis.

Q: What is the most common cause of viral conjunctivitis?

A: Adenovirus (certain serotypes, but don't be concerned with which ones)

Q: Where does the poliovirus replicate during initial infection?

A: Although it can have serious CNS manifestations, the poliovirus is an enterovirus and replicates in the GI tract.

Q: Which of the hepatitis viruses are transmitted by the fecal–oral route?

A: Hepatitis A and E. Remember this hint? List the five hepatitis viruses in order: A, B, C, D, and E. A and E are on the *outside* of the other three viruses. Hepatitis A and E come from *outside* (contaminated food/water), thus fecal–oral. The three remaining viruses are B, C, and D and are on the *inside* of A and E. Blood is on the *inside* of the body; therefore these three are blood-borne.

Q: How would you classify a person with the following hepatitis serology: HBsAg negative, anti-HBe positive, anti-Hbc positive, anti-HBs negative, HBAg negative?

A: Infected in the past

Q: How would you distinguish a person who is immune to hepatitis B as a result of a past infection from one who is immune after a vaccination?

A: Both anti-HBs and anti-Hbc are seen in the serum after an infection, whereas only anti-HBs is seen after immunization.

Q: What is the name of the papular syphilitic lesions near the genital area in secondary syphilis?

A: Condyloma latum. Condylomata acuminata are genital warts (venereal warts) due to papillomaviruses. The latter are frequently cauliflower shaped.

Q: Describe a positive Tzanck test.

A: Epidermal multinucleated giant cells (keratinocytes) are seen. They are characteristic of herpes virus infections.

Q: What is the name of the mechanism by which *Trypanosoma brucei* evades the immune system?

A: Antigenic variation. It does so by activating a gene ("switching") that codes for a new surface coat glycoprotein to which the immune system has not been exposed.

Q: What drugs are recommended for prophylaxis for *Pneumocystis jiroveci* (*carinii*) pneumonia?

A: Oral trimethoprim-sulfamethoxazole is the first choice. Other drugs include dapsone, dapsone + pyrimethamine + leucovorin, atovaquone, or aerosolized pentamidine. Therapy can be discontinued for patients receiving highly active antiretroviral therapy with an increase in CD4+ count to more than 200 cells/mm^3 for at least three months.

Q: What hypersensitivity reaction produces many erythematous plaques with dusky centers and red borders resembling targets?

A: Erythema multiforme. This disease can also produce nonpruritic urticarial lesions, petechiae, vesicles, and bullae. The severe form is called Stevens-Johnson syndrome.

Q: What can cause erythema multiforme?

A: Another long list of viruses (e.g., herpes simplex virus, EBV, Coxsackie, varicella-zoster virus, mumps), bacteria (e.g., *Mycoplasma pneumoniae, Brucella, Corynebacterium diphtheriae, Yersinia, Mycobacterium, Neisseria gonorrhoeae*), protozoa, drugs (penicillins, sulfonamides, salicylates, anticonvulsants, barbiturates), pregnancy, premenstrual hormone changes, and malignancy (the cause is unknown in half the cases)

Q: What can cause erythema nodosum?

A: As in the previous question, another long list of viruses, bacteria, fungi, drugs, pregnancy, and malignancy

Q: What is the number one cause of aseptic meningitis?

A: Enteroviruses, in general, such as echoviruses and Coxsackievirus.

Q: Where in the United States is coccidioidomycosis most prevalent?

A: The semiarid Southwest (Lower Sonoran life zone)

Q: Currant jelly sputum is indicative of pneumonia due to what organism?

A: *Klebsiella pneumoniae*. This organism has a large mucoid capsule. Because the pneumonia is necrotic, the sputum can also contain blood.

Q: A 21-year-old man presents with a dry cough, malaise, fever, and a sore throat that developed in the past two weeks. What is the diagnosis?

A: Pneumonia due to *Mycoplasma pneumoniae* is most likely. This disease usually has a slow onset and occurs primarily in the teen years and in young adults. Treat the patient with erythromycin or tetracycline.

Q: What is the most frequent cause of nosocomial pneumonia?

A: *Pseudomonas aeruginosa*. It usually occurs in immunocompromised patients or patients on mechanical ventilation.

Q: A 56-year-old smoker comes to your office after a three-day business convention he attended the previous week. He is nauseated, coughing, and has chills, fever of 103.5°F, and pulse of 68. Should you be concerned about your own health after he coughs in your direction?

A: No. This patient probably has Legionnaires' disease. *Legionella pneumophila* contaminates the water in air conditioning towers and other moist environments. It is not transmitted from person-to-person.

Q: What are the classic chest x-ray findings in a patient with mycoplasma pneumonia?

A: Patchy diffuse densities involving the entire lung

Q: A 43-year-old man presents with pleurisy, sudden onset of fever and chills, and rust-colored sputum. What is the diagnosis?

A: Pneumococcal pneumonia caused by *Streptococcus pneumoniae*, the most common community-acquired pneumonia. It is a consolidating lobar pneumonia.

Q: Describe the different presentations of bacterial and viral pneumonia.

A: Bacterial pneumonia is typified by a sudden onset of symptoms, including pleurisy, fever and chills, productive cough, tachypnea, and tachycardia. The most common bacterial pneumonia is due to *S. pneumoniae*. Viral pneumonia is characterized by a gradual onset of symptoms, general malaise, no pleurisy, chills or high fever, and a nonproductive cough.

Q: What part of an antibody molecule binds complement?

A: The Fc region of IgG (not IgG$_4$) or IgM

Q: What secondary bacterial pneumonias often occur after a viral pneumonia?

A: Pneumonia due to *Staphylococcus aureus, Strep. pneumoniae*, and *H. influenzae* are some.

Q: Which type of pneumonia is often associated with cold agglutinins?

A: *Mycoplasma pneumoniae*. Detection of cold agglutinins can be useful in diagnosis, but it is nonspecific. Newer serological tests, such as a rapid enzyme-linked immunosorbent assay, are available and usually preferable to the older tests.

Q: What are two major risk factors for the development of anaerobic lung infection?

A: Periodontal disease and predisposition to aspiration

Q: Which bacterial pneumonias frequently cause frank hemoptysis?

A: *Pseudomonas aeruginosa*, *Klebsiella pneumoniae*, and *Staphylococcus aureus*. Hemoptysis can have other infectious and noninfectious etiologies. Up to 70% are due to infections.

Q: The recovery of *Pseudomonas aeruginosa*, particularly the mucoid form, from the lower respiratory tract of a child or young adult with chronic lung symptoms is common in what disease?

A: Cystic fibrosis

Q: A 17-year-old boy complains of dysphagia and fever. Tonsillar exudates and anterior cervical adenitis are noted. What sign differentiates *Streptococcus pyogenes* from other causes of pharyngitis?

A: A sandpaper-like rash, especially in the skinfolds, such as under the arms, in the groin, and on the buttocks

Q: A 24-year-old man complains that he has endured two days of watery stools, muscle cramps, and extreme fatigue. He looks pale, dehydrated, and very ill. The patient states that he has just returned from India. What is the diagnosis?

A: Most likely cholera

Q: Which diarrheal illnesses cause fecal leukocytes?

A: *Shigella*, *Campylobacter*, and enteroinvasive *E. coli* are three of the most common. Others include *Salmonella*, *Yersinia enterocolitica*, *Vibrio parahaemolyticus*, and *C. difficile*. Fecal white blood cells are absent in toxigenic and enteropathogenic infection, even with such a virulent organism as *Vibrio cholerae*. Viral and parasitic infections rarely produce fecal white blood cells.

Q: What group of organisms is the most common cause of acute diarrhea?

A: Viruses. Viral diarrhea is generally self-limited, lasting only 1–3 days.

Q: A patient with new diarrhea and abdominal pain has been on antibiotics for eight days for an unrelated infection. What might be revealed by sigmoidoscopy?

A: Yellowish superficial plaques. This finding is indicative of pseudomembranous colitis. Stool studies should be assayed for the exotoxins of *Clostridium difficile*.

Q: If a particular patient were at increased risk of respiratory infections because of a specific immunoglobulin deficiency, what immunoglobulin would most likely be deficient?

A: IgA. Secretory IgA can be found in various secretions, including mucus. It functions as part of our first line of defense in prevention of colonization and attachment of various microbes.

Q: A 20-year-old college student is home for winter break and presents with a 10-day history of a nonproductive dry hacking cough, malaise, mild fever, and no chills. What is the diagnosis?

A: *Mycoplasma pneumoniae*, also known as walking pneumonia. Although this is one of the most common pneumonias that develop in teenagers and young adults, it is an atypical pneumonia and most frequently occurs in close contact populations (i.e., schools and military barracks). There are other less common causes of walking pneumonia such as *Chlamydophila pneumoniae*, *Legionella pneumophila* (Pontiac fever), and *Coxiella burnetii* (Q fever).

Q: What is the drug of choice for pseudomembranous colitis?

A: Oral metronidazole. Vancomycin is an alternate.

Q: A patient presents with acute meningitis. When should antibiotics be initiated?

A: Immediately. Acute meningitis is a medical emergency. Empiric treatment should include ceftriaxone or cefotaxime plus vancomycin. The two most common causes of bacterial meningitis are *Streptococcus pneumoniae* and *Neisseria meningitidis*.

Q: In the United States what animals are most likely to be infected with the rabies virus?

A: Bats, skunks, and raccoons. Dogs are the usual carriers in developing countries. Cases due to dog bites in the United States are rare.

Q: What is tabes dorsalis?

A: Progressive loss of all or part of the body's reflexes. The large joints of affected limbs are destroyed. Patients experience severe stabbing pains in their legs, sensory deficits, and difficulty walking. Forty percent of patients with neurosyphilis are afflicted with tabes dorsalis.

Q: What are the clinical and laboratory features of Reye syndrome?

A: Viral illness followed by vomiting, behavioral changes, hepatomegaly, hyperventilation, and coma

Q: What is the medical treatment of confirmed botulism?

A: Trivalent (anti-A, -B, and -E) botulism antitoxin and respiratory support. The mortality rate is typically 100% after symptoms begin, although there are rare exceptions.

Q: What is the most common cause of focal encephalitis in AIDS patients?

A: Toxoplasmosis. Symptoms include focal neurological deficits, headache, fever, altered mental status, and seizures. The disease is much less common in immunocompetent adults.

Q: What is the cause of chancroid?

A: *Haemophilus ducreyi*. Patients with this condition present with one or more painful necrotic lesions. A painful genital ulcer in combination with tender suppurative inguinal adenopathy suggests chancroid (soft chancre). It is common in tropical areas but rare elsewhere.

Q: What is the most common tapeworm in North America?

A: *Hymenolepis nana*. It probably is the most common tapeworm worldwide. This small tapeworm (dwarf tapeworm <4 cm) commonly infects children, especially in areas with poor sanitation. Humans are infected by ingestion of the egg, most commonly by the direct fecal–oral route. Infection is usually asymptomatic, unless there is a high parasite load. It is the only human tapeworm that has no intermediate host. Praziquantel is the drug of choice. In the United States infection is most common in the Southeast (<1% of children).

Q: A patient presents with fever, dyspnea, cough, hemoptysis, and eosinophilia. What is the likely diagnosis?

A: Infection with *Ascaris lumbricoides*. The worms can migrate to the lungs, resulting in pulmonary manifestations. As with most parasitic infections, eosinophilia is common. Infection can be treated with mebendazole, pyrantel pamoate, or albendazole.

Q: How is infection with the hookworm *Necator americanus* or *Ancylostoma duodenale* acquired?

A: The larvae in infected soil can penetrate intact skin, especially of those walking barefoot on infected soil. The disease is most common in warm climates, including the southern United States. Patients present with chronic anemia (iron deficiency), cough, low-grade fever, diarrhea, abdominal pain, weakness, weight loss, eosinophilia, and guaiac positive stools. A diagnosis is confirmed if ova are present in the stools. Treatment includes mebendazole, pyrantel pamoate, or albendazole. Of the two species of hookworms, *Necator americanus* is most common in the United States.

Q: What are the signs and symptoms of *Trichuris trichiura* infection?

A: The adult of this nematode (whipworm) lives in the cecum. Larvae penetrate the mucosa of the large intestine where they mature into adults. Symptoms include anorexia, abdominal pain (especially right upper quadrant), insomnia, fever, diarrhea, flatulence, weight loss, pruritus, eosinophilia, and microcytic hypochromic anemia. Rectal prolapse is a severe manifestation. Most infections are in children. The diagnosis is made by microscopic detection of the barrel-shaped eggs, with polar plugs at each end, in the stools. Mebendazole is the drug of choice.

Q: What are three common protozoa that can cause diarrhea?

A: *Entamoeba histolytica*: Found worldwide; most cases in the United States are in immigrants. Although half of the infected patients are asymptomatic, the usual symptoms consist of nausea, vomiting, diarrhea, flatulence, anorexia, abdominal pain, and leukocytosis. Diagnosis is based on observation of cysts in stools and serology. Treatment is with metronidazole or tinidazole.

Giardia duodenalis: Found worldwide. This organism is one of the most common intestinal parasites in the United States. Symptoms include explosive watery diarrhea, flatulence, abdominal distention, fatigue, and fever. The diagnosis is confirmed via a stool examination. Treatment is with metronidazole.

Cryptosporidium parvum: Found worldwide. Symptoms are profuse watery diarrhea, cramps, nausea, vomiting, flatulence, and weight loss. The infection is usually self-limiting in immunocompetent individuals, but it can be treated with nitazoxanide. This drug is not as effective in the immunosuppressed.

Q: What is the most deadly form of malaria?

A: Malaria due to *Plasmodium falciparum*

Q: How is *P. falciparum* diagnosed on blood smear?

A: Observation of (1) small ring forms with double chromatin knobs within the erythrocyte, (2) multiple rings infected within red blood cells, (3) rare trophozoites and schizonts on smear, (4) pathognomonic crescent-shaped gametocytes, and (5) parasitemia exceeding 4%.

Q: What is the drug of choice for treating malaria?
A: Chloroquine, except for resistant strains of *P. falciparum* and *P. vivax* (combination therapy with quinine and doxycycline or quinine and pyrimethamine sulfadoxine)

Q: What parasitic infection commonly results in a papular pruritic rash?
A: Schistosomiasis. The rash is a hypersensitivity reaction (allergic dermatitis) to antigens of *Schistosoma*. Systemic infection occurs when the cercariae enter the body through the skin, travel through the circulatory system, and develop into mature worms in the portal veins.

Q: Infections with what group of parasites does not typically result in eosinophilia?
A: Protozoan infections, such as with amebas, *Giardia*, *Trypanosoma*, and *Babesia*

Q: Name the most common intestinal parasite in the United States.
A: *Giardia duodenalis* (syn. *G. lamblia* or *G. intestinalis*)

Q: How is Chagas disease transmitted?
A: Primarily by the blood-sucking reduviid, or "kissing" bug. The parasite is in the feces of the bug and can enter the body while the bug is feeding. The disease is also known as American trypanosomiasis and is due to *Trypanosoma cruzi*.

Q: What three diseases are transmitted by the deer tick, *Ixodes scapularis*?
A: Lyme disease, human granulocytic anaplasmosis, and babesiosis in the northeastern United States

Q: How do patients present with *Babesia* infection?
A: Intermittent fever, splenomegaly, jaundice, and hemolysis. The disease may be fatal in patients without spleens. Questions on this disease are not common on standardized exams.

Q: What causes Colorado tick fever?
A: A reovirus (Colorado tick fever virus). The vector is the wood tick, *Dermacentor andersoni*. The disease is self-limiting; treatment is supportive with acetaminophen for the fever and pain. It has an acute onset and lasts 5–10 days. In about half of the cases there is a first phase with fever lasting 2–3 days, followed by a 1- to 3-day period without fever but with anorexia and malaise. A second phase consists of a return of the fever and an increase in symptoms that last for about 48 hours. The two staged fever and a recent tick bite are important clues.

Q: *Streptococcus pyogenes* and *Streptococcus agalactiae* are the two most common streptococci that are routinely beta-hemolytic. How are these two differentiated from each other?
A: *Streptococcus pyogenes* (group A strep) is bacitracin sensitive. It is also positive in the PYR (pyrrolidyl arylamidase) test, but this test is less likely to appear on a standardized examination. In addition, *S. agalactiae* (group B strep) is positive in the CAMP test and it can hydrolyze hippurate.

Q: What is the most common cause of cellulitis?
A: *Streptococcus pyogenes*. *Staphylococcus aureus* is also a common cause of cellulitis, although it is generally less severe and more often associated with an open wound. *Haemophilus influenzae* type b is a cause of cellulitis in young children. Other bacteria are occasional causes.

Q: What is the most common cause of cutaneous abscesses?
A: *Staphylococcus aureus*

Q: A 26-year-old woman presents with headache, fever, malaise, and tender regional lymphadenopathy about a week after a cat bite. A tender papule develops at the site. What is the diagnosis?
A: Cat-scratch disease. This condition usually develops three days to six weeks after a cat bite or scratch. The papule typically blisters and heals with eschar formation. A transient macular or vesicular rash may also develop.

Q: What causes cat-scratch disease?
A: *Bartonella henselae*

Q: What is the most common cause of gas gangrene?
A: *Clostridium perfringens*, but about 20 other species of clostridia can cause the infection.
Trivia: Fournier gangrene is a polymicrobial infection (severe necrotizing fasciitis) of the scrotum. It is usually caused by anaerobes other than the clostridia. (This is a bonus question; it is doubtful it will be on a standardized exam.)

Q: What is the most common site of a herpes simplex 1 infection?

A: The lower lip. These lesions are painful and can frequently recur because the virus remains in the trigeminal ganglia. Stress, sun, and illness generally trigger recurrences.

Q: What are the sequelae of viral meningitis?

A: Typically none

Q: At which stage of Lyme disease does neurological involvement occur?

A: The second and third stages. Second stage: cranial neuropathies, meningitis, and radiculoneuritis. Third stage: encephalitis and a variety of CNS manifestations, including stroke-like syndromes, extrapyramidal, and cerebellar involvement

Q: What virus is responsible for subacute sclerosing panencephalitis?

A: The measles virus

Q: To which group of viruses does the poliovirus belong?

A: Poliovirus is an enterovirus that belongs to the picornavirus group.

Q: Where is the location of the P24 antigen in HIV?

A: It is a capsid protein found in the core of the virion.

Q: A patient is diagnosed with impetigo from *Streptococcus pyogenes*. What poststreptococcal sequelae might develop?

A: Acute poststreptococcal glomerulonephritis. Rheumatic fever is more common after pharyngitis.

Q: A patient presents to the hospital with fever, chills, and leukocytosis one month after placement of a mechanical prosthetic valve. Endocarditis is suspected. What bacteria are most common?

A: *Staphylococcus aureus* or *Staphylococcus epidermidis* are most common early. Viridans streptococci as well as *Staphylococcus epidermidis* and *S. aureus* are seen in late disease.

Q: What is the drug of choice for *Neisseria meningitidis* infection?

A: Penicillin G

Q: What prophylactic antibiotics should be given to people who have had contact with patients with meningococcal meningitis?

A: Rifampin or a fluoroquinolone are recommended.

Q: What is the most common cause of aseptic meningitis?

A: The enteroviruses

Q: What is the mechanism of action of the aminoglycosides?

A: They irreversibly bind to the 30S subunit of 70S ribosomes. This causes an inhibition of protein synthesis by two mechanisms: (1) misreading of the triplet codon of mRNA, causing the wrong amino acid to be inserted into the growing peptide, and (2) the premature release of the growing peptide chain. Both of these can result in aberrant (nonfunctional) proteins.

Q: What is the recommended drug of choice for gonorrhea?

A: Ceftriaxone. Other choices include cefixime, cefotaxime, and penicillin G. Azithromycin or doxycycline should also be given because of the high incidence of coinfection with *Chlamydia trachomatis*.

Q: What is the cause of epidemic keratoconjunctivitis?

A: Adenovirus. As the name of the disease implies, it is highly contagious.

Q: What is the drug of choice for streptococcal pharyngitis?

A: Penicillin

Q: What is the most common cause of herpangina?

A: Coxsackie A virus, usually, but also coxsackievirus B, echovirus, and enterovirus. It usually occurs in children in the summer.

Q: What is the most common cause of diarrhea in a child younger than 1 year old?

A: Rotavirus

Q: Why does therapy for TB take months, when other infections usually clear in a matter of days?

A: The mycobacteria divide very slowly (long generation time) and have a long dormant phase, during which time they are not responsive to medications. Treatment is typically 9–12 months.

Q: What causes erythema nodosum?

A: Erythema nodosum is a hypersensitivity reaction that is associated with many infectious and some noninfectious processes. Some of its better-known associations are *Streptococcus pyogenes*, *Neisseria meningitidis*, *Treponema pallidum*, *Mycobacterium tuberculosis*, and *Mycobacterium leprae* as well as diseases such as histoplasmosis, coccidioidomycosis, blastomycosis, and herpes simplex virus infection. The red painful nodules are usually on the lower legs but can be on the arms or face.

Q: What are the most common organisms found in human bite wounds?

A: *Staphylococcus aureus*, *Streptococcus* species, and *Eikenella corrodens*. Anaerobes are also commonly seen. Infections are usually polymicrobic.

Q: What major bacterial organism should be considered in a dog bite?

A: *Pasteurella multocida*. Other bacteria, including *Staphylococcus aureus*, *Streptococcus*, *E. coli*, and anaerobes, can be involved. About 20% of dog bites become infected. The most serious are those that involve bones, joints, or tendons; severe hand bites; and cranial bites in infants.

Q: A patient in Kansas presents with diarrhea, high fever, headache, lethargy, and confusion. CSF values are normal, 45% band forms are seen on the differential of his white blood count, and a blood culture is positive for *Escherichia coli*. What is the most likely cause of the diarrhea?

A: Tough question! *Shigella sonnei* is the most likely pathogen. Blood cultures in *Shigella* diarrhea are virtually never positive for *Shigella*. When they are positive, they are more likely to be positive for *E. coli*. This might be because although *Shigella* is locally quite invasive at the mucosal level, it is poorly invasive at the systemic level. Resident *E. coli* in the gut, however, take advantage of the disrupted mucosa and invade the blood stream. *S. sonnei* is the most common species in Kansas and the rest of the United States.

Q: What is the mode of action of daptomycin and what is its clinical use?

A: Daptomycin is a lipopeptide with a spectrum of activity limited to gram-positive organisms, including methicillin-resistant *Staphylococcus aureus*, vancomycin-resistant *S. aureus*, and vancomycin-resistant *Enterococcus*. It is FDA approved for the treatment of skin and soft tissue infections. It causes disruption of the bacterial membrane through the formation of transmembrane channels, resulting in leakage of intracellular ions. The result is depolarization of the cellular membrane and inhibition of macromolecular synthesis.

Q: You are a physician driving home on Tuesday after working in your rural (e.g., not in any of the large cities) California clinic and pass a dead squirrel (on the side of the road, it is not road kill). On Wednesday, taking a different route, you pass two more dead squirrels, again, not road kill. The following morning in your clinic you see a 26-year-old man with enlarged tender lymphadenitis and a 105°F fever. What illness should be highly suspected in your differential that would be of minimal concern had you not seen the dead squirrels?

A: Plague. Cases of human plague (*Yersinia pestis*) are sometimes preceded by squirrel deaths. A squirrel die-off occurs when the organism is introduced into a highly susceptible mammalian population, causing a high mortality rate among infected animals. This is referred to as epizootic plague when the epidemic is in animals. An epidemic of plague among humans is called urban plague and does not occur in the United States.

Q: A patient from the Philippines has a hypopigmented patch that is lacking in sensation. What is the most likely cause of his problem?

A: Hansen disease (leprosy) due to *Mycobacterium leprae*. Note: A dermatologist in Nebraska recently relayed that she has diagnosed about six cases of lepromatous leprosy, mostly in international travelers, in her 30-year career.

Q: What is the major animal reservoir for *Leptospira interrogans* in the United States?

A: Dogs, but rats would also be a good guess

Q: Which intestinal parasites are known to cause anemia as a major manifestation?

A: Hookworms

Q: Which class of immunoglobulins is responsible for urticaria (hives)?

A: IgE

Q: Which class of immunoglobulins is responsible for food allergies?

A: IgE

Q: How long after exposure to an allergen does anaphylaxis occur?

A: Seconds to one hour

Q: What lymphocyte surface molecule is responsible for HLA class II antigens?
    A: CD4

Q: What are the only complement fixing immunoglobulins?
    A: IgG and IgM

Q: What is the major immunoglobulin that is protective on mucous membranes?
    A: IgA

Q: Which immunoglobulin is the major host defense against parasites?
    A: IgE

Q: What are the two main functions of T cells?
    A: To signal B cells to make antibody and to kill virally infected or tumor cells

Q: Briefly explain the cellular basis for the type I hypersensitivity reaction (wheal and flare). Give a clinical example.
    A: This immediate type or anaphylactic hypersensitivity is mediated by circulating basophils and mast cells, which become activated by cross-linking of IgE on their membrane surface. The prototypic IgE mediated disease is ragweed hay fever. Other, sometimes fatal, anaphylactic reactions are the classic insect venom or food-induced allergies.

Q: Briefly explain the cellular basis for the type II hypersensitivity reaction (cytotoxic). Give a clinical example.
    A: These immune interactions involve integral cellular antigen components and IgG and IgM antibody formation to these foreign antigen determinants. The classic example is immune mediated hemolysis such as that seen in transfusion reactions or hemolytic disease of the newborn.

Q: Briefly explain the cellular basis for the type III hypersensitivity reaction (Arthus or immune complex). Give a clinical example.
    A: Tissue injury is caused by immune complex deposition in various tissues, which are toxic to that tissue, by mechanisms, such as complement activation or proteolytic enzyme release. Examples include immune complex pericarditis and arthritis after meningococcal or *H. influenzae* infection.

Q: In type IV hypersensitivity reaction (cell mediated or delayed type), pathological changes follow interaction of antigen with what cellular component of the immune system?
    A: Antigen-specific sensitized T cells

Q: Which hypersensitivity reaction is responsible for most glomerulopathies?
    A: Type III or immune complex disease

Q: What is a prototypic delayed-type hypersensitivity type IV reaction?
    A: Contact allergy, such as chemical-induced contact dermatitis or poison ivy

Q: Patients with severe allergic reaction to eggs should avoid receipt of which vaccines?
    A: Influenza and yellow fever vaccines

Q: What viruses are associated with posttransplantation malignancy?
    A: Lymphoma due to EBV infection is common. Other viruses include HHV-8, hepatitis C virus, and hepatitis B virus.

Q: What is the most sensitive and specific sign of an infectious disease in an immunocompromised host?
    A: Fever

Q: Reiter syndrome may occur after infection with what microbial agents?
    A: *Chlamydia trachomatis* is the most common, but it can also occur after infections with *Shigella*, *Yersinia enterocolitica*, and *Campylobacter*.

Q: There is simultaneous activation of coagulation and fibrinolysis in what pathological condition?
    A: Disseminated intravascular coagulation (DIC)

Q: What is the predominant symptom in DIC?
    A: Bleeding, because the clotting factors are depleted

Q: A patient presents with a raised, red, small, and painful plaque on the face. Upon examination, a distinct sharp advancing edge is noted. What is the cause?

A: This is most likely erysipelas, which is caused by *Streptococcus pyogenes*. Although the face is commonly involved, it can affect other areas of the body.

Q: What is a furuncle and its most common cause?

A: A deep-inflammatory nodule that grows out of superficial folliculitis (i.e., it is a boil) due to *S. aureus*.

Q: A father brought his 6-year-old daughter to the clinic because of a rash on her face that has patchy areas with vesiculopustular lesions covered in a thick honey-colored crust. Just two days ago these lesions were small red papules. What is your diagnosis?

A: Impetigo. This is most common in children and usually occurs on exposed areas of skin.

Q: What organism is responsible for the above patient's infection?

A: It depends on what you read or whom you ask. It is either *Streptococcus pyogenes* or *Staphylococcus aureus*. Impetigo due to *S. pyogenes* resembles the "honey crust" described above and *S. aureus* is frequently bullous (fluid-filled vesicles).

Q: A 72-year-old female has a painful red rash with crops of blisters on erythematous bases in a band-like distribution on the right side of her lower back, which spreads down and out toward her hip. What is your diagnosis?

A: Shingles or herpes zoster. This is due to the reactivation of dormant varicella virus in the sensory root ganglia of a patient with a history of chickenpox. The rash is in the distribution of the dermatome, in this case, L5. It is most common in the elderly population or in patients who are immunocompromised. Treatment is with acyclovir and oral analgesics.

Q: Can shingles be prevented?

A: A qualified yes. The FDA-approved vaccine given at age 60 can prevent about 50% of cases.

Q: What is the infectious cause of scalded skin syndrome?

A: *Staphylococcus aureus*. Drugs or chemicals can also be a cause. Both begin with the appearance of patches of tender erythema followed by loosening of the skin and denuding to glistening bases. Staphylococcal scalded skin syndrome is commonly found in children younger than 5 years. It is due to a toxin (exfoliatin) that cleaves within the epidermis under the stratum granulosum.

Q: What anatomical areas are usually affected by staphylococcal scalded skin syndrome?

A: The face around nose and mouth, neck, axillae, and groin. The disease commonly occurs after upper respiratory tract infections or purulent conjunctivitis. Nikolsky sign (lateral pressure on the skin results in epidermal separation from the dermis) is positive.

Q: What is the most common cause of Stevens-Johnson syndrome?

A: Drugs, especially sulfa drugs. Other causes are responses to infections with *Mycoplasma pneumoniae* and herpes simplex virus. The disease is self-limiting but severely uncomfortable.

Q: What is the treatment of a tetanus prone wound?

A: Surgical debridement, intramuscular administration of human tetanus immune globulin, and metronidazole (drug of choice), penicillin G, or doxycycline

Q: What is tinea capitis?

A: A fungal infection of the scalp that begins as a papule around one hair shaft and then spreads to other follicles. The infection can cause the hairs to break off, leaving little black dot stumps and patches of alopecia. *Trichophyton* and *Microsporum* species cause most cases in North America. Wood's lamp (UV light) examination will cause fluorescence of only *Microsporum*. KOH preparations can be used to visualize the hyphae. The infection is also called "ringworm of the scalp." Treatment is systemic with griseofulvin, itraconazole, or terbinafine.

Q: What organisms most commonly cause tinea pedis and onychomycosis?

A: *Trichophyton* species. Tinea pedis is more commonly called athletes foot, and onychomycosis is a fungal infection of the nails.

Q: What is the causative agent of pityriasis versicolor (formerly tinea versicolor)?

A: *Malassezia furfur* (*Pityrosporum orbiculare*), a fungus. It presents as hypopigmented areas especially in summer because they do not tan normally. A KOH scraping of an affected area reveals abundant short hyphae and spores ("spaghetti and meatballs"). It most often involves the upper trunk.

Q: An elderly patient with chronic obstructive pulmonary disease is most likely to contract what kind of pneumonia?
   A: Pneumonias due to *Streptococcus pneumoniae*, *Haemophilus influenzae*, or *Moraxella catarrhalis* are common. This population is at risk and should be vaccinated with PPV23 and Hib.

Q: What is the most common cause of community-acquired pneumonia in the elderly?
   A: *Streptococcus pneumoniae*, *Haemophilus influenzae*, and *Staphylococcus aureus* (in that order) are most common. *Chlamydophila pneumoniae* is also seen.

Q: What is the most common cause of UTIs in uncatheterized elderly patients?
   A: *E. coli*

Q: What is the mechanism of action of the fluoroquinolones?
   A: They are bactericidal via inhibition of DNA gyrase.

Q: A 27-year-old woman presents with a maculopapular rash on her palms and soles. She complains of headaches and general weakness. On examination, you find she has multiple condylomata lata and lymphadenopathy. What is the diagnosis?
   A: Secondary syphilis. This develops 6–9 weeks after the syphilitic chancre, which will have resolved by this time. Treatment is with penicillin G.

Q: A 22-year-old man, who has no significant medical history and is taking no medication, presents with a creamy white coat on his tongue. The substance easily rubs off, revealing an erythematous base. What medical condition should be considered?
   A: HIV. In a patient who has no obvious reason for having an overgrowth of oral *Candida*, HIV should be suspected. Other causes for oral thrush include cancer, systemic illness, neutropenia, diabetes, adrenal insufficiency, nutritional deficiencies, or an immunocompromised state.

Q: What are the classic symptoms of TB?
   A: Night sweats, fever, weight loss, malaise, cough, and greenish-yellow sputum most commonly seen in the mornings.

Q: What do chest x-rays reveal in cases of tuberculosis?
   A: Cavitation of the right upper lobe. Lower lung infiltrates, hilar adenopathy, atelectasis, and pleural effusion are also common.

Q: What does an elevated IgM anti-HBc indicate?
   A: Exposure to hepatitis B with antibody to the core antigen. High titers indicate the contagious disease; low titers suggest chronic hepatitis B.

Q: Which type of hepatitis is caused by a DNA virus?
   A: Hepatitis B. The incubation period is 90 days.

Q: What does an anti-HB indicate?
   A: Prior infection and immunity

Q: What is the Waterhouse-Friderichsen syndrome?
   A: The Waterhouse-Friderichsen syndrome is the name applied to bilateral adrenal hemorrhage that develops as a sequela of infectious disease. It is most often observed as a consequence of septicemia secondary to meningococcemia. The patient typically has a petechial rash, purpura, shaking, chills, and a severe headache. Septic shock is imminent. The syndrome can also be associated with other diseases such as pneumonia due to community-acquired methicillin-resistant *Staphylococcus aureus* (CA-MRSA) in children and some *S. pneumoniae* infections. Patients can survive with rapid and aggressive therapy including antibiotics.

Q: What is the most common type of hepatitis transmitted through blood transfusions?
   A: Hepatitis C

Q: A woman comes to your office frantic because her husband has just received a positive VDRL result. They have been happily married for 35 years and she cannot believe he has been unfaithful. Is it at all possible that he has been loyal to his wife?
   A: Yes. False-positive tests can occur if the patient has had a viral or mycoplasma infection in the near past, if the patient is an intravenous drug user, or if the patient has systemic lupus erythematosus. The presence or absence of syphilis can be confirmed with the FTA-absorption or other tests.

Q: How is hepatitis A transmitted?
>A: By the oral–fecal route. No carrier state exists.

Q: Which type of bacterial gastroenteritis involves diarrhea and is associated with the consumption of seafood?
>A: *Vibrio parahaemolyticus* (most commonly), a halophile

Q: What is the cause of condylomata acuminata?
>A: Papilloma virus

Q: What does the India ink test show?
>A: Capsules of bacteria and of *Cryptococcus neoformans*

Q: In patients with bacterial meningitis, is intracranial pressure usually elevated?
>A: Yes, usually

Q: An 8-year-old boy has recently acquired a pet iguana. This may place the household at increased risk for what disease?
>A: Salmonellosis. Reptiles can be carriers of *Salmonella*.

Q: Of the three bacteria listed (*Salmonella*, *Shigella*, and *Campylobacter*), which one, when ingested in contaminated food, typically requires the *least* number of cells to cause food poisoning?
>A: *Shigella*. Because *Shigella* is more resistant to the low pH of the stomach, fewer organisms (50–200) are required to cause disease.

Q: From the previous question, which one of the three food-borne bacterial pathogens listed is found only in humans (humans serve as the reservoir)?
>A: *Shigella*

Q: What inorganic ions do siderophores bind?
>A: Iron. Pyoverdin is the greenish fluorescent pigment of *Pseudomonas aeruginosa*. It is a siderophore and appears to be essential for virulence.

Q: What is an alternative name for a penicillin binding protein?
>A: Transpeptidase

Q: A 7-month-old male with a fever of 104°F was taken to the emergency department by his parents. After an evaluation, the resident on duty suspected bacterial meningitis. CSF was obtained and sent to the microbiology lab for culture. What media should be used for culturing?
>A: At a minimum, a blood agar plate and chocolate agar (supplemented with X and V factors) should be inoculated. At this age, the two most important causes of meningitis are *Streptococcus pneumoniae* and *Haemophilus influenzae*, but never discount other causes. The supplemented chocolate agar is for the isolation and growth of *H. influenzae*.

Q: List some infections that involve biofilms.
>A: Almost anything you mention, except virus and maybe parasite infections. Most infections involve biofilm formation. Some of the most obvious or notorious include dental caries, gingivitis, periodontitis, infections involving prosthetic and indwelling devices (e.g., intravascular catheters, cerebrospinal fluid shunts, prosthetic cardiac valves, etc.), UTIs, endocarditis, and cystic fibrosis. We can probably assume any bacterial, yeast, and fungal infection involves biofilm formation unless proven otherwise.

Q: What are the three major methods of genetic transfer between bacteria?
>A: Transformation, conjugation, and transduction

Q: In bacterial genetics, what is transformation?
>A: It is the uptake by bacteria of free DNA strands from the local environment. This transfer can be inhibited with the enzyme DNase that will digest the DNA before it can enter the recipient cell.

Q: In bacterial genetics, what is conjugation?
>A: It is the transfer of DNA between two cells, a donor/male/F⁺ and a recipient/female/F⁻. It requires cell-to-cell contact. DNase does not inhibit this transfer because the DNA is always protected within the cells during the transfer. Conjugation can be inhibited if the F⁺ and F⁻ cells are separated by a membrane filter with pores that are too small (about 0.45 μm or less) to allow the bacteria to pass through the filter. The filter will not stop transformation or transduction.

Q: In bacterial genetics, what is transduction?

A: It is the transfer of DNA from one cell to another via a bacterial virus (bacteriophage). This transfer is not inhibited by DNase and not inhibited if the cells are separated by a membrane filter with pores of about 0.45 μm or less (although there are some filters with pores which are small enough to prevent the passage of viruses).

Q: Name some aminoglycosides.

A: Gentamicin, kanamycin, tobramycin, amikacin, and streptomycin are several

Q: What is the antibacterial spectrum of the aminoglycosides?

A: There are three important things to remember about the aminoglycosides. First, they are most active against gram-negative bacteria; second, both gram-positive and -negative anaerobes (as well as the genus *Streptococcus*) are resistant; and third, they are bactericidal.

Q: What is the mode of action of the aminoglycosides?

A: They irreversibly bind (thus they are bactericidal) to the A-site of the 16S rRNA on the 30S ribosomal subunit. This has two effects: the mRNA is misread (the wrong amino acid is inserted) and the growing peptide chain is prematurely released. Both of these can result in aberrant (nonfunctional) proteins.

Q: What are the mechanisms of resistance to the aminoglycosides?

A: As with many antimicrobials, resistance due to the cells becoming less permeable (decreased uptake/accumulation) is always a good guess. In addition, resistance can be more specific due to the bacteria producing an enzyme (there are several different ones) that modifies the drug so that there is decreased binding to the ribosomes and usually decreased uptake. A third mechanism is a mutation that modifies the ribosomal binding site, thus preventing the aminoglycoside from binding (this mechanism is most significant for streptomycin).

Q: A person with sickle-cell anemia is at an increased risk of infection with what organisms?

A: Damage to the spleen makes them susceptible to encapsulated bacteria such as *Streptococcus pneumoniae*, *Haemophilus influenzae* type b, and *Neisseria meningitidis*. Other impairments to the immune system cause increased susceptibility to other pathogens, such as *Salmonella* (osteomyelitis), *Staphylococcus aureus*, *E. coli*, and *Klebsiella*. Human parvovirus B19 can cause a temporary pause in erythrocyte production (aplastic crisis).

Q: A person with sickle-cell anemia is at a decreased risk of infection with what organism?

A: *Plasmodium* species, the etiological agents of malaria

Q: Are fermentation and anaerobic respiration the same?

A: No. Fermentation just involves the conversion of pyruvic acid into fermentation byproducts. No energy is produced. Anaerobic respiration uses some of the cytochromes, but not all, to generate energy in the form of ATP. In anaerobic respiration, some other inorganic ion other than oxygen (such as nitrate or sulfate) serves as the terminal electron acceptor. Not as much ATP is produced in anaerobic respiration as in aerobic respiration, where oxygen serves as terminal electron acceptor.

Q: What is Weil disease?

A: It refers to severe cases of leptospirosis (jaundice, vasculitis, kidney damage, myocarditis, etc.).

Q: A bioterrorist could choose botulism as his disease of choice. How would the disease be disseminated?

A: He would use the toxin, not the organism. The toxin can be put in food, but it is 100 times more toxic if inhaled. Aerosolization would be the delivery method of choice. The symptoms are the same as food-borne botulism.

Q: How is West Nile fever diagnosed?

A: Detection of IgG and IgM antibodies to West Nile virus in serum or CSF

Q: What is the significance of IgM antibody in CSF?

A: Because IgM antibody does not cross the blood–brain barrier, IgM antibody in CSF strongly suggests CNS infection.

Q: What causes severe acute respiratory syndrome, or SARS?

A: SARS coronavirus (SARS-CoV). As of this writing, the last cases (two lab workers) were in China in 2004.

Q: What kind of virus causes avian flu?

A: It is an influenza A virus (an orthomyxovirus), serotype H_5N_1.

Q: Did the avian influenza virus appear because of an antigenic drift or shift?

A: Shift. It was due to the reassortment of the segments of the RNA genome when a host (possibly a hog) was coinfected with two different influenza viruses. In 1997, the virus was transferred directly from birds to humans. The virus is highly virulent with a death rate over 60%.

Q: Is there an avian influenza vaccine?

A: Yes. The FDA approved of a vaccine in 2007 to be used by public health authorities on a need basis.